Rainwater Analysis

STEM Road Map
for Elementary School

Grade
5

Rainwater Analysis

Grade 5

STEM Road Map
for Elementary School

Edited by Carla C. Johnson, Janet B. Walton, and
Erin Peters-Burton

National Science Teachers Association

Arlington, Virginia

National Science Teachers Association

Claire Reinburg, Director
Rachel Ledbetter, Managing Editor
Andrea Silen, Associate Editor
Jennifer Thompson, Associate Editor
Donna Yudkin, Book Acquisitions Manager

ART AND DESIGN
Will Thomas Jr., Director, cover and
  interior design
Himabindu Bichali, Graphic Designer, interior
  design

PRINTING AND PRODUCTION
Catherine Lorrain, Director

NATIONAL SCIENCE TEACHERS ASSOCIATION
David L. Evans, Executive Director

1840 Wilson Blvd., Arlington, VA 22201
www.nsta.org/store
For customer service inquiries, please call 800-277-5300.

NSTA is committed to publishing material that promotes the best in inquiry-based science education. However, conditions of actual use may vary, and the safety procedures and practices described in this book are intended to serve only as a guide. Additional precautionary measures may be required. NSTA and the authors do not warrant or represent that the procedures and practices in this book meet any safety code or standard of federal, state, or local regulations. NSTA and the authors disclaim any liability for personal injury or damage to property arising out of or relating to the use of this book, including any of the recommendations, instructions, or materials contained therein.

PERMISSIONS

Library of Congress Cataloging-in-Publication Data
Names: Johnson, Carla C., 1969- editor. | Walton, Janet B., 1968- editor. | Peters-Burton, Erin E., editor. | National Science Teachers Association, issuing body.
Title: Rainwater analysis / edited by Carla C. Johnson, Janet B. Walton, and Erin Peters-Burton.
Description: Arlington, VA : National Science Teachers Association, 2018. | Series: STEM road map for elementary school. Grade 5 | Includes bibliographical references and index.
Identifiers: LCCN 2018040694 (print) | LCCN 2018051171 (ebook) | ISBN 9781681404509 (e-book) | ISBN 9781681404493 (print)
Subjects: LCSH: Water harvesting--Study and teaching (Elementary) | Rain and rainfall--Study and teaching (Elementary) | Gardens--Irrigation--Study and teaching (Elementary)
Classification: LCC TD418 (ebook) | LCC TD418 .R357 2018 (print) | DDC 628.1/1--dc23
LC record available at *https://lccn.loc.gov/2018040694*

The *Next Generation Science Standards* ("NGSS") were developed by twenty-six states, in collaboration with the National Research Council, the National Science Teachers Association and the American Association for the Advancement of Science in a process managed by Achieve, Inc. For more information go to *www.nextgenscience.org.*

CONTENTS

CONTENTS

ABOUT THE EDITORS AND AUTHORS

Dr. Carla C. Johnson is the associate dean for research, engagement, and global partnerships and a professor of science education at Purdue University's College of Education in West Lafayette, Indiana. Dr. Johnson serves as the director of research and evaluation for the Department of Defense–funded Army Educational Outreach Program (AEOP), a global portfolio of STEM education programs, competitions, and apprenticeships. She has been a leader in STEM education for the past decade, serving as the director of STEM Centers, editor of the *School Science and Mathematics* journal, and lead researcher for the evaluation of Tennessee's Race to the Top–funded STEM portfolio. Dr. Johnson has published over 100 articles, books, book chapters, and curriculum books focused on STEM education. She is a former science and social studies teacher and was the recipient of the 2013 Outstanding Science Teacher Educator of the Year award from the Association for Science Teacher Education (ASTE), the 2012 Award for Excellence in Integrating Science and Mathematics from the School Science and Mathematics Association (SSMA), the 2014 award for best paper on Implications of Research for Educational Practice from ASTE, and the 2006 Outstanding Early Career Scholar Award from SSMA. Her research focuses on STEM education policy implementation, effective science teaching, and integrated STEM approaches.

Dr. Janet B. Walton is a research assistant professor and the assistant director of evaluation for AEOP at Purdue University's College of Education. Formerly the STEM workforce program manager for Virginia's Region 2000 and founding director of the Future Focus Foundation, a nonprofit organization dedicated to enhancing the quality of STEM education in the region, she merges her economic development and education backgrounds to develop K–12 curricular materials that integrate real-life issues with sound cross-curricular content. Her research focuses on collaboration between schools and community stakeholders for STEM education and problem- and project-based learning pedagogies. With this research agenda, she works to forge productive relationships between K–12 schools and local business and community stakeholders to bring contextual STEM experiences into the classroom and provide students and educators with innovative resources and curricular materials.

Dr. Erin Peters-Burton is the Donna R. and David E. Sterling endowed professor in science education at George Mason University in Fairfax, Virginia. She uses her experiences from 15 years as an engineer and secondary science, engineering, and mathematics teacher to develop research projects that directly inform classroom practice in science and engineering. Her research agenda is based on the idea that all students should build self-awareness of how they learn science and engineering. She works to help students see themselves as "science-minded" and help teachers create classrooms that support student skills to develop scientific knowledge. To accomplish this, she pursues research projects that investigate ways that students and teachers can use self-regulated learning theory in science and engineering, as well as how inclusive STEM schools can help students succeed. During her tenure as a secondary teacher, she had a National Board Certification in Early Adolescent Science and was an Albert Einstein Distinguished Educator Fellow for NASA. As a researcher, Dr. Peters-Burton has published over 100 articles, books, book chapters, and curriculum books focused on STEM education and educational psychology. She received the Outstanding Science Teacher Educator of the Year award from ASTE in 2016 and a Teacher of Distinction Award and a Scholarly Achievement Award from George Mason University in 2012, and in 2010 she was named University Science Educator of the Year by the Virginia Association of Science Teachers.

Dr. Tamara J. Moore is an associate professor of engineering education in the College of Engineering at Purdue University. Dr. Moore's research focuses on defining STEM integration through the use of engineering as the connection and investigating its power for student learning.

Paula Schoeff taught grades K–8 for 20 years in Indiana and Ohio and assisted in helping educate teachers in Ohio and Kentucky in STEM practices. Schoeff has a master's degree from the University of Cincinnati in curriculum and instruction with a focus on science.

Dr. Toni A. Sondergeld is an associate professor of assessment, research, and statistics in the School of Education at Drexel University in Philadelphia. Dr. Sondergeld's research concentrates on assessment and evaluation in education, with a focus on K–12 STEM.

ACKNOWLEDGMENTS

This module was developed as a part of the STEM Road Map project (Carla C. Johnson, principal investigator). The Purdue University College of Education, General Motors, and other sources provided funding for this project.

See *www.routledge.com/products/9781138804234* for more information about *STEM Road Map: A Framework for Integrated STEM Education.*

PART 1

THE STEM ROAD MAP

BACKGROUND, THEORY, AND PRACTICE

OVERVIEW OF THE *STEM ROAD MAP CURRICULUM SERIES*

Carla C. Johnson, Erin Peters-Burton, and Tamara J. Moore

The *STEM Road Map Curriculum Series* was conceptualized and developed by a team of STEM educators from across the United States in response to a growing need to infuse real-world learning contexts, delivered through authentic problem-solving pedagogy, into K–12 classrooms. The curriculum series is grounded in integrated STEM, which focuses on the integration of the STEM disciplines—science, technology, engineering, and mathematics—delivered across content areas, incorporating the Framework for 21st Century Learning along with grade-level-appropriate academic standards.

The curriculum series begins in kindergarten, with a five-week instructional sequence that introduces students to the STEM themes and gives them grade-level-appropriate topics and real-world challenges or problems to solve. The series uses project-based and problem-based learning, presenting students with the problem or challenge during the first lesson, and then teaching them science, social studies, English language arts, mathematics, and other content, as they apply what they learn to the challenge or problem at hand.

Authentic assessment and differentiation are embedded throughout the modules. Each *STEM Road Map Curriculum Series* module has a lead discipline, which may be science, social studies, English language arts, or mathematics. All disciplines are integrated into each module, along with ties to engineering. Another key component is the use of STEM Research Notebooks to allow students to track their own learning progress. The modules are designed with a scaffolded approach, with increasingly complex concepts and skills introduced as students progress through grade levels.

The developers of this work view the curriculum as a resource that is intended to be used either as a whole or in part to meet the needs of districts, schools, and teachers who are implementing an integrated STEM approach. A variety of implementation formats are possible, from using one stand-alone module at a given grade level to using all five modules to provide 25 weeks of instruction. Also, within each grade band (K–2, 3–5, 6–8, 9–12), the modules can be sequenced in various ways to suit specific needs.

STANDARDS-BASED APPROACH

The *STEM Road Map Curriculum Series* is anchored in the *Next Generation Science Standards (NGSS)*, the *Common Core State Standards for Mathematics (CCSS Mathematics)*, the *Common Core State Standards for English Language Arts (CCSS ELA)*, and the Framework for 21st Century Learning. Each module includes a detailed curriculum map that incorporates the associated standards from the particular area correlated to lesson plans. The STEM Road Map has very clear and strong connections to these academic standards, and each of the grade-level topics was derived from the mapping of the standards to ensure alignment among topics, challenges or problems, and the required academic standards for students. Therefore, the curriculum series takes a standards-based approach and is designed to provide authentic contexts for application of required knowledge and skills.

THEMES IN THE *STEM ROAD MAP CURRICULUM SERIES*

The K–12 STEM Road Map is organized around five real-world STEM themes that were generated through an examination of the big ideas and challenges for society included in STEM standards and those that are persistent dilemmas for current and future generations:

- Cause and Effect
- Innovation and Progress
- The Represented World
- Sustainable Systems
- Optimizing the Human Experience

These themes are designed as springboards for launching students into an exploration of real-world learning situated within big ideas. Most important, the five STEM Road Map themes serve as a framework for scaffolding STEM learning across the K–12 continuum.

The themes are distributed across the STEM disciplines so that they represent the big ideas in science (Cause and Effect; Sustainable Systems), technology (Innovation and Progress; Optimizing the Human Experience), engineering (Innovation and Progress; Sustainable Systems; Optimizing the Human Experience), and mathematics (The Represented World), as well as concepts and challenges in social studies and 21st century skills that are also excellent contexts for learning in English language arts. The process of developing themes began with the clustering of the *NGSS* performance expectations and the National Academy of Engineering's grand challenges for engineering, which led to the development of the challenge in each module and connections of the module activities to the *CCSS Mathematics* and *CCSS ELA* standards. We performed these

mapping processes with large teams of experts and found that these five themes provided breadth, depth, and coherence to frame a high-quality STEM learning experience from kindergarten through 12th grade.

Cause and Effect

The concept of cause and effect is a powerful and pervasive notion in the STEM fields. It is the foundation of understanding how and why things happen as they do. Humans spend considerable effort and resources trying to understand the causes and effects of natural and designed phenomena to gain better control over events and the environment and to be prepared to react appropriately. Equipped with the knowledge of a specific cause-and-effect relationship, we can lead better lives or contribute to the community by altering the cause, leading to a different effect. For example, if a person recognizes that irresponsible energy consumption leads to global climate change, that person can act to remedy his or her contribution to the situation. Although cause and effect is a core idea in the STEM fields, it can actually be difficult to determine. Students should be capable of understanding not only when evidence points to cause and effect but also when evidence points to relationships but not direct causality. The major goal of education is to foster students to be empowered, analytic thinkers, capable of thinking through complex processes to make important decisions. Understanding causality, as well as when it cannot be determined, will help students become better consumers, global citizens, and community members.

Innovation and Progress

One of the most important factors in determining whether humans will have a positive future is innovation. Innovation is the driving force behind progress, which helps create possibilities that did not exist before. Innovation and progress are creative entities, but in the STEM fields, they are anchored by evidence and logic, and they use established concepts to move the STEM fields forward. In creating something new, students must consider what is already known in the STEM fields and apply this knowledge appropriately. When we innovate, we create value that was not there previously and create new conditions and possibilities for even more innovations. Students should consider how their innovations might affect progress and use their STEM thinking to change current human burdens to benefits. For example, if we develop more efficient cars that use by-products from another manufacturing industry, such as food processing, then we have used waste productively and reduced the need for the waste to be hauled away, an indirect benefit of the innovation.

The Represented World

When we communicate about the world we live in, how the world works, and how we can meet the needs of humans, sometimes we can use the actual phenomena to explain a concept. Sometimes, however, the concept is too big, too slow, too small, too fast, or too complex for us to explain using the actual phenomena, and we must use a representation or a model to help communicate the important features. We need representations and models such as graphs, tables, mathematical expressions, and diagrams because it makes our thinking visible. For example, when examining geologic time, we cannot actually observe the passage of such large chunks of time, so we create a timeline or a model that uses a proportional scale to visually illustrate how much time has passed for different eras. Another example may be something too complex for students at a particular grade level, such as explaining the p subshell orbitals of electrons to fifth graders. Instead, we use the Bohr model, which more closely represents the orbiting of planets and is accessible to fifth graders.

When we create models, they are helpful because they point out the most important features of a phenomenon. We also create representations of the world with mathematical functions, which help us change parameters to suit the situation. Creating representations of a phenomenon engages students because they are able to identify the important features of that phenomenon and communicate them directly. But because models are estimates of a phenomenon, they leave out some of the details, so it is important for students to evaluate their usefulness as well as their shortcomings.

Sustainable Systems

From an engineering perspective, the term *system* refers to the use of "concepts of component need, component interaction, systems interaction, and feedback. The interaction of subcomponents to produce a functional system is a common lens used by all engineering disciplines for understanding, analysis, and design." (Koehler, Bloom, and Binns 2013, p. 8). Systems can be either open (e.g., an ecosystem) or closed (e.g., a car battery). Ideally, a system should be sustainable, able to maintain equilibrium without much energy from outside the structure. Looking at a garden, we see flowers blooming, weeds sprouting, insects buzzing, and various forms of life living within its boundaries. This is an example of an ecosystem, a collection of living organisms that survive together, functioning as a system. The interaction of the organisms within the system and the influences of the environment (e.g., water, sunlight) can maintain the system for a period of time, thus demonstrating its ability to endure. Sustainability is a desirable feature of a system because it allows for existence of the entity in the long term.

In the STEM Road Map project, we identified different standards that we consider to be oriented toward systems that students should know and understand in the K–12 setting. These include ecosystems, the rock cycle, Earth processes (such as erosion, tectonics, ocean currents, weather phenomena), Earth-Sun-Moon cycles, heat transfer, and the interaction among the geosphere, biosphere, hydrosphere, and atmosphere. Students and teachers should understand that we live in a world of systems that are not independent of each other, but rather are intrinsically linked such that a disruption in one part of a system will have reverberating effects on other parts of the system.

Optimizing the Human Experience

Science, technology, engineering, and mathematics as disciplines have the capacity to continuously improve the ways humans live, interact, and find meaning in the world, thus working to optimize the human experience. This idea has two components: being more suited to our environment and being more fully human. For example, the progression of STEM ideas can help humans create solutions to complex problems, such as improving ways to access water sources, designing energy sources with minimal impact on our environment, developing new ways of communication and expression, and building efficient shelters. STEM ideas can also provide access to the secrets and wonders of nature. Learning in STEM requires students to think logically and systematically, which is a way of knowing the world that is markedly different from knowing the world as an artist. When students can employ various ways of knowing and understand when it is appropriate to use a different way of knowing or integrate ways of knowing, they are fully experiencing the best of what it is to be human. The problem-based learning scenarios provided in the STEM Road Map help students develop ways of thinking like STEM professionals as they ask questions and design solutions. They learn to optimize the human experience by innovating improvements in the designed world in which they live.

THE NEED FOR AN INTEGRATED STEM APPROACH

At a basic level, STEM stands for science, technology, engineering, and mathematics. Over the past decade, however, STEM has evolved to have a much broader scope and broader implications. Now, educators and policy makers refer to STEM as not only a concentrated area for investing in the future of the United States and other nations but also as a domain and mechanism for educational reform.

The good intentions of the recent decade-plus of focus on accountability and increased testing has resulted in significant decreases not only in instructional time for teaching science and social studies but also in the flexibility of teachers to promote authentic, problem solving–focused classroom environments. The shift has had a detrimental impact on student acquisition of vitally important skills, which many refer to as 21st century skills, and often the ability of students to "think." Further, schooling has become increasingly siloed into compartments of mathematics, science, English language arts, and social studies, lacking any of the connections that are overwhelmingly present in

the real world around children. Students have experienced school as content provided in boxes that must be memorized, devoid of any real-world context, and often have little understanding of why they are learning these things.

STEM-focused projects, curriculum, activities, and schools have emerged as a means to address these challenges. However, most of these efforts have continued to focus on the individual STEM disciplines (predominantly science and engineering) through more STEM classes and after-school programs in a "STEM enhanced" approach (Breiner et al. 2012). But in traditional and STEM enhanced approaches, there is little to no focus on other disciplines that are integral to the context of STEM in the real world. Integrated STEM education, on the other hand, infuses the learning of important STEM content and concepts with a much-needed emphasis on 21st century skills and a problem- and project-based pedagogy that more closely mirrors the real-world setting for society's challenges. It incorporates social studies, English language arts, and the arts as pivotal and necessary (Johnson 2013; Rennie, Venville, and Wallace 2012; Roehrig et al. 2012).

FRAMEWORK FOR STEM INTEGRATION IN THE CLASSROOM

The *STEM Road Map Curriculum Series* is grounded in the Framework for STEM Integration in the Classroom as conceptualized by Moore, Guzey, and Brown (2014) and Moore et al. (2014). The framework has six elements, described in the context of how they are used in the *STEM Road Map Curriculum Series* as follows:

1. The STEM Road Map contexts are meaningful to students and provide motivation to engage with the content. Together, these allow students to have different ways to enter into the challenge.

2. The STEM Road Map modules include engineering design that allows students to design technologies (i.e., products that are part of the designed world) for a compelling purpose.

3. The STEM Road Map modules provide students with the opportunities to learn from failure and redesign based on the lessons learned.

4. The STEM Road Map modules include standards-based disciplinary content as the learning objectives.

5. The STEM Road Map modules include student-centered pedagogies that allow students to grapple with the content, tie their ideas to the context, and learn to think for themselves as they deepen their conceptual knowledge.

6. The STEM Road Map modules emphasize 21st century skills and, in particular, highlight communication and teamwork.

All of the STEM Road Map modules incorporate these six elements; however, the level of emphasis on each of these elements varies based on the challenge or problem in each module.

THE NEED FOR THE *STEM ROAD MAP CURRICULUM SERIES*

As focus is increasing on integrated STEM, and additional schools and programs decide to move their curriculum and instruction in this direction, there is a need for high-quality, research-based curriculum designed with integrated STEM at the core. Several good resources are available to help teachers infuse engineering or more STEM enhanced approaches, but no curriculum exists that spans K–12 with an integrated STEM focus. The next chapter provides detailed information about the specific pedagogy, instructional strategies, and learning theory on which the *STEM Road Map Curriculum Series* is grounded.

REFERENCES

Breiner, J., M. Harkness, C. C. Johnson, and C. Koehler. 2012. What is STEM? A discussion about conceptions of STEM in education and partnerships. *School Science and Mathematics* 112 (1): 3–11.

Johnson, C. C. 2013. Conceptualizing integrated STEM education: Editorial. *School Science and Mathematics* 113 (8): 367–368.

Koehler, C. M., M. A. Bloom, and I. C. Binns. 2013. Lights, camera, action: Developing a methodology to document mainstream films' portrayal of nature of science and scientific inquiry. *Electronic Journal of Science Education* 17 (2).

Moore, T. J., S. S. Guzey, and A. Brown. 2014. Greenhouse design to increase habitable land: An engineering unit. *Science Scope* 51–57.

Moore, T. J., M. S. Stohlmann, H.-H. Wang, K. M. Tank, A. W. Glancy, and G. H. Roehrig. 2014. Implementation and integration of engineering in K–12 STEM education. In *Engineering in pre-college settings: Synthesizing research, policy, and practices,* ed. S. Purzer, J. Strobel, and M. Cardella, 35–60. West Lafayette, IN: Purdue Press.

Rennie, L., G. Venville, and J. Wallace. 2012. *Integrating science, technology, engineering, and mathematics: Issues, reflections, and ways forward.* New York: Routledge.

Roehrig, G. H., T. J. Moore, H. H. Wang, and M. S. Park. 2012. Is adding the *E* enough? Investigating the impact of K–12 engineering standards on the implementation of STEM integration. *School Science and Mathematics* 112 (1): 31–44.

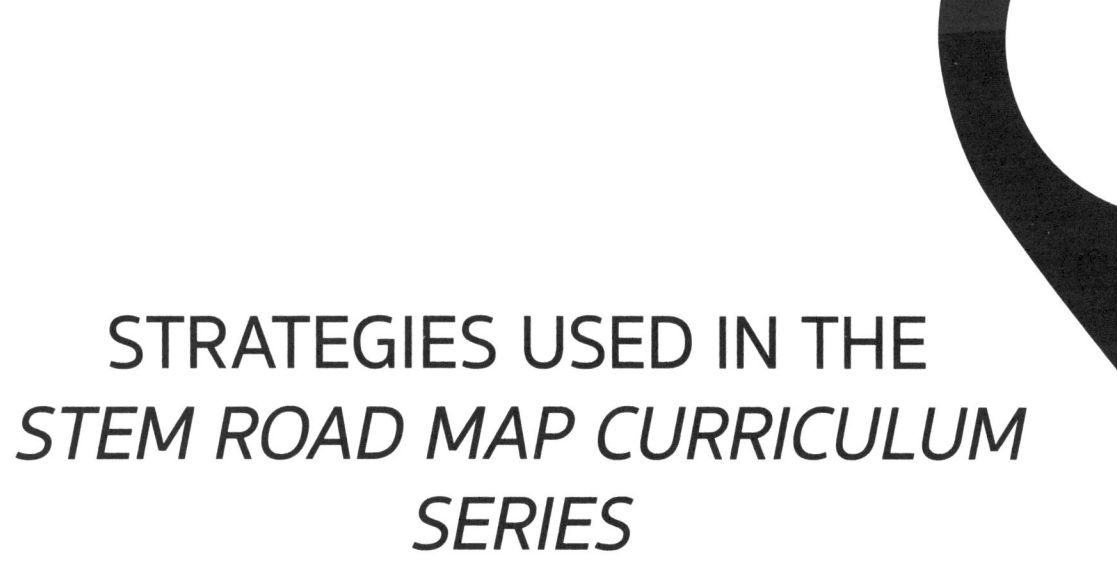

STRATEGIES USED IN THE *STEM ROAD MAP CURRICULUM SERIES*

Erin Peters-Burton, Carla C. Johnson, Toni A. Sondergeld, and Tamara J. Moore

The *STEM Road Map Curriculum Series* uses what has been identified through research as best-practice pedagogy, including embedded formative assessment strategies throughout each module. This chapter briefly describes the key strategies that are employed in the series.

PROJECT- AND PROBLEM-BASED LEARNING

Each module in the *STEM Road Map Curriculum Series* uses either project-based learning or problem-based learning to drive the instruction. Project-based learning begins with a driving question to guide student teams in addressing a contextualized local or community problem or issue. The outcome of project-based instruction is a product that is conceptualized, designed, and tested through a series of scaffolded learning experiences (Blumenfeld et al. 1991; Krajcik and Blumenfeld 2006). Problem-based learning is often grounded in a fictitious scenario, challenge, or problem (Barell 2006; Lambros 2004). On the first day of instruction within the unit, student teams are provided with the context of the problem. Teams work through a series of activities and use open-ended research to develop their potential solution to the problem or challenge, which need not be a tangible product (Johnson 2003).

ENGINEERING DESIGN PROCESS

The *STEM Road Map Curriculum Series* uses engineering design as a way to facilitate integrated STEM within the modules. The engineering design process (EDP) is depicted in Figure 2.1 (p. 10). It highlights two major aspects of engineering design—problem scoping and solution generation—and six specific components of working toward a design: define the problem, learn about the problem, plan a solution, try the solution, test the solution, decide whether the solution is good enough. It also shows that communication

Figure 2.1. Engineering Design Process

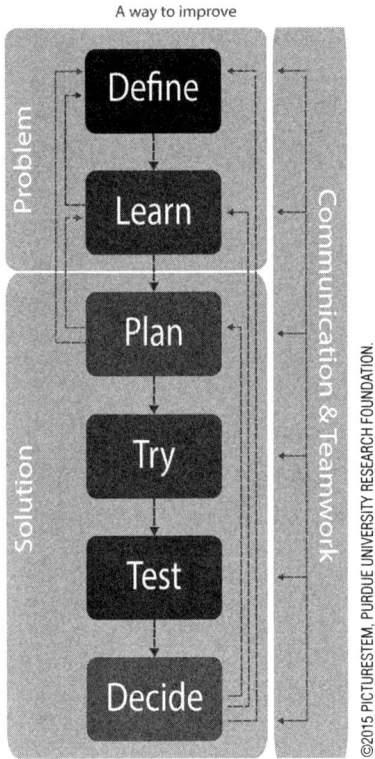

and teamwork are involved throughout the entire process. As the arrows in the figure indicate, the order in which the components of engineering design are addressed depends on what becomes needed as designers progress through the EDP. Designers must communicate and work in teams throughout the process. The EDP is iterative, meaning that components of the process can be repeated as needed until the design is good enough to present to the client as a potential solution to the problem.

Problem scoping is the process of gathering and analyzing information to deeply understand the engineering design problem. It includes defining the problem and learning about the problem. Defining the problem includes identifying the problem, the client, and the end user of the design. The client is the person (or people) who hired the designers to do the work, and the end user is the person (or people) who will use the final design. The designers must also identify the criteria and the constraints of the problem. The criteria are the things the client wants from the solution, and the constraints are the things that limit the possible solutions. The designers must spend significant time learning about the problem, which can include activities such as the following:

- Reading informational texts and researching about relevant concepts or contexts

- Identifying and learning about needed mathematical and scientific skills, knowledge, and tools

- Learning about things done previously to solve similar problems

- Experimenting with possible materials that could be used in the design

Problem scoping also allows designers to consider how to measure the success of the design in addressing specific criteria and staying within the constraints over multiple iterations of solution generation.

Solution generation includes planning a solution, trying the solution, testing the solution, and deciding whether the solution is good enough. Planning the solution includes generating many design ideas that both address the criteria and meet the constraints. Here the designers must consider what was learned about the problem during problem scoping. Design plans include clear communication of design ideas through media such as notebooks, blueprints, schematics, or storyboards. They also include details about the design, such as measurements, materials, colors, costs of materials, instructions for how things fit together, and sets of directions. Making the decision about which design idea to move forward involves considering the trade-offs of each design idea.

Once a clear design plan is in place, the designers must try the solution. Trying the solution includes developing a prototype (a testable model) based on the plan generated. The prototype might be something physical or a process to accomplish a goal. This component of design requires that the designers consider the risk involved in implementing the design. The prototype developed must be tested. Testing the solution includes conducting fair tests that verify whether the plan is a solution that is good enough to meet the client and end user needs and wants. Data need to be collected about the results of the tests of the prototype, and these data should be used to make evidence-based decisions regarding the design choices made in the plan. Here, the designers must again consider the criteria and constraints for the problem.

Using the data gathered from the testing, the designers must decide whether the solution is good enough to meet the client and end user needs and wants by assessment based on the criteria and constraints. Here, the designers must justify or reject design decisions based on the background research gathered while learning about the problem and on the evidence gathered during the testing of the solution. The designers must now decide whether to present the current solution to the client as a possibility or to do more iterations of design on the solution. If they decide that improvements need to be made to the solution, the designers must decide if there is more that needs to be understood about the problem, client, or end user; if another design idea should be tried; or if more planning needs to be conducted on the same design. One way or another, more work needs to be done.

Throughout the process of designing a solution to meet a client's needs and wants, designers work in teams and must communicate to each other, the client, and likely the end user. Teamwork is important in engineering design because multiple perspectives and differing skills and knowledge are valuable when working to solve problems. Communication is key to the success of the designed solution. Designers must communicate their ideas clearly using many different representations, such as text in an engineering notebook, diagrams, flowcharts, technical briefs, or memos to the client.

LEARNING CYCLE

The same format for the learning cycle is used in all grade levels throughout the STEM Road Map, so that students engage in a variety of activities to learn about phenomena in the modules thoroughly and have consistent experiences in the problem- and project-based learning modules. Expectations for learning by younger students are not as high as for older students, but the format of the progression of learning is the same. Students who have learned with curriculum from the STEM Road Map in early grades know

what to expect in later grades. The learning cycle consists of five parts—Introductory Activity/Engagement, Activity/Exploration, Explanation, Elaboration/Application of Knowledge, and Evaluation/Assessment—and is based on the empirically tested 5E model from BSCS (Bybee et al. 2006).

In the Introductory Activity/Engagement phase, teachers introduce the module challenge and use a unique approach designed to pique students' curiosity. This phase gets students to start thinking about what they already know about the topic and begin wondering about key ideas. The Introductory Activity/Engagement phase positions students to be confident about what they are about to learn, because they have prior knowledge, and clues them into what they don't yet know.

In the Activity/Exploration phase, the teacher sets up activities in which students experience a deeper look at the topics that were introduced earlier. Students engage in the activities and generate new questions or consider possibilities using preliminary investigations. Students work independently, in small groups, and in whole-group settings to conduct investigations, resulting in common experiences about the topic and skills involved in the real-world activities. Teachers can assess students' development of concepts and skills based on the common experiences during this phase.

During the Explanation phase, teachers direct students' attention to concepts they need to understand and skills they need to possess to accomplish the challenge. Students participate in activities to demonstrate their knowledge and skills to this point, and teachers can pinpoint gaps in student knowledge during this phase.

In the Elaboration/Application of Knowledge phase, teachers present students with activities that engage in higher-order thinking to create depth and breadth of student knowledge, while connecting ideas across topics within and across STEM. Students apply what they have learned thus far in the module to a new context or elaborate on what they have learned about the topic to a deeper level of detail.

In the last phase, Evaluation/Assessment, teachers give students summative feedback on their knowledge and skills as demonstrated through the challenge. This is not the only point of assessment (as discussed in the section on Embedded Formative Assessments), but it is an assessment of the culmination of the knowledge and skills for the module. Students demonstrate their cognitive growth at this point and reflect on how far they have come since the beginning of the module. The challenges are designed to be multidimensional in the ways students must collaborate and communicate their new knowledge.

STEM RESEARCH NOTEBOOK

One of the main components of the *STEM Road Map Curriculum Series* is the STEM Research Notebook, a place for students to capture their ideas, questions, observations, reflections, evidence of progress, and other items associated with their daily work. At the beginning of each module, the teacher walks students through the setup of the STEM

Research Notebook, which could be a three-ring binder, composition book, or spiral notebook. You may wish to have students create divided sections so that they can easily access work from various disciplines during the module. Electronic notebooks kept on student devices are also acceptable and encouraged. Students will develop their own table of contents and create chapters in the notebook for each module.

Each lesson in the *STEM Road Map Curriculum Series* includes one or more prompts that are designed for inclusion in the STEM Research Notebook and appear as questions or statements that the teacher assigns to students. These prompts require students to apply what they have learned across the lesson to solve the big problem or challenge for that module. Each lesson is designed to meaningfully refer students to the larger problem or challenge they have been assigned to solve with their teams. The STEM Research Notebook is designed to be a key formative assessment tool, as students' daily entries provide evidence of what they are learning. The notebook can be used as a mechanism for dialogue between the teacher and students, as well as for peer and self-evaluation.

The use of the STEM Research Notebook is designed to scaffold student notebooking skills across the grade bands in the *STEM Road Map Curriculum Series*. In the early grades, children learn how to organize their daily work in the notebook as a way to collect their products for future reference. In elementary school, students structure their notebooks to integrate background research along with their daily work and lesson prompts. In the upper grades (middle and high school), students expand their use of research and data gathering through team discussions to more closely mirror the work of STEM experts in the real world.

THE ROLE OF ASSESSMENT IN THE *STEM ROAD MAP CURRICULUM SERIES*

Starting in the middle years and continuing into secondary education, the word *assessment* typically brings grades to mind. These grades may take the form of a letter or a percentage, but they typically are used as a representation of a student's content mastery. If well thought out and implemented, however, classroom assessment can offer teachers, parents, and students valuable information about student learning and misconceptions that does not necessarily come in the form of a grade (Popham 2013).

The *STEM Road Map Curriculum Series* provides a set of assessments for each module. Teachers are encouraged to use assessment information for more than just assigning grades to students. Instead, assessments of activities requiring students to actively engage in their learning, such as student journaling in STEM Research Notebooks, collaborative presentations, and constructing graphic organizers, should be used to move student learning forward. Whereas other curriculum with assessments may include objective-type (multiple-choice or matching) tests, quizzes, or worksheets, we have intentionally avoided these forms of assessments to better align assessment strategies with teacher instruction and

student learning techniques. Since the focus of this book is on project- or problem-based STEM curriculum and instruction that focuses on higher-level thinking skills, appropriate and authentic performance assessments were developed to elicit the most reliable and valid indication of growth in student abilities (Brookhart and Nitko 2008).

Comprehensive Assessment System

Assessment throughout all STEM Road Map curriculum modules acts as a comprehensive system in which formative and summative assessments work together to provide teachers with high-quality information on student learning. Formative assessment occurs when the teacher finds out formally or informally what a student knows about a smaller, defined concept or skill and provides timely feedback to the student about his or her level of proficiency. Summative assessments occur when students have performed all activities in the module and are given a cumulative performance evaluation in which they demonstrate their growth in learning.

A comprehensive assessment system can be thought of as akin to a sporting event. Formative assessments are the practices: It is important to accomplish them consistently, they provide feedback to help students improve their learning, and making mistakes can be worthwhile if students are given an opportunity to learn from them. Summative assessments are the competitions: Students need to be prepared to perform at the best of their ability. Without multiple opportunities to practice skills along the way through formative assessments, students will not have the best chance of demonstrating growth in abilities through summative assessments (Black and Wiliam 1998).

Embedded Formative Assessments

Formative assessments in this module serve two main purposes: to provide feedback to students about their learning and to provide important information for the teacher to inform immediate instructional needs. Providing feedback to students is particularly important when conducting problem- or project-based learning because students take on much of the responsibility for learning, and teachers must facilitate student learning in an informed way. For example, if students are required to conduct research for the Activity/Exploration phase but are not familiar with what constitutes a reliable resource, they may develop misconceptions based on poor information. When a teacher monitors this learning through formative assessments and provides specific feedback related to the instructional goals, students are less likely to develop incomplete or incorrect conceptions in their independent investigations. By using formative assessment to detect problems in student learning and then acting on this information, teachers help move student learning forward through these teachable moments.

Formative assessments come in a variety of formats. They can be informal, such as asking students probing questions related to student knowledge or tasks or simply observing students engaged in an activity to gather information about student skills. Formative assessments can also be formal, such as a written quiz or a laboratory practical. Regardless of the type, three key steps must be completed when using formative assessments (Sondergeld, Bell, and Leusner 2010). First, the assessment is delivered to students so that teachers can collect data. Next, teachers analyze the data (student responses) to determine student strengths and areas that need additional support. Finally, teachers use the results from information collected to modify lessons and create learning environments that reinforce weak points in student learning. If student learning information is not used to modify instruction, the assessment cannot be considered formative in nature.

Formative assessments can be about content, science process skills, or even learning skills. When a formative assessment focuses on content, it assesses student knowledge about the disciplinary core ideas from the *Next Generation Science Standards* (*NGSS*) or content objectives from *Common Core State Standards for Mathematics* (*CCSS Mathematics*) or *Common Core State Standards for English Language Arts* (*CCSS ELA*). Content-focused formative assessments ask students questions about declarative knowledge regarding the concepts they have been learning. Process skills formative assessments examine the extent to which a student can perform science and engineering practices from the *NGSS* or process objectives from *CCSS Mathematics* or *CCSS ELA*, such as constructing an argument. Learning skills can also be assessed formatively by asking students to reflect on the ways they learn best during a module and identify ways they could have learned more.

Assessment Maps

Assessment maps or blueprints can be used to ensure alignment between classroom instruction and assessment. If what students are learning in the classroom is not the same as the content on which they are assessed, the resultant judgment made on student learning will be invalid (Brookhart and Nitko 2008). Therefore, the issue of instruction and assessment alignment is critical. The assessment map for this book (found in Chapter 3) indicates by lesson whether the assessment should be completed as a group or on an individual basis, identifies the assessment as formative or summative in nature, and aligns the assessment with its corresponding learning objectives.

Note that the module includes far more formative assessments than summative assessments. This is done intentionally to provide students with multiple opportunities to practice their learning of new skills before completing a summative assessment. Note also that formative assessments are used to collect information on only one or two learning objectives at a time so that potential relearning or instructional modifications can focus on smaller and more manageable chunks of information. Conversely, summative assessments in the module cover many more learning objectives, as they are traditionally used as final markers of student learning. This is not to say that information collected from summative assessments cannot or should not be used formatively. If teachers find that gaps in student

learning persist after a summative assessment is completed, it is important to revisit these existing misconceptions or areas of weakness before moving on (Black et al. 2003).

SELF-REGULATED LEARNING THEORY IN THE STEM ROAD MAP MODULES

Many learning theories are compatible with the STEM Road Map modules, such as constructivism, situated cognition, and meaningful learning. However, we feel that the self-regulated learning theory (SRL) aligns most appropriately (Zimmerman 2000). SRL requires students to understand that thinking needs to be motivated and managed (Ritchhart, Church, and Morrison 2011). The STEM Road Map modules are student centered and are designed to provide students with choices, concrete hands-on experiences, and opportunities to see and make connections, especially across subjects (Eliason and Jenkins 2012; NAEYC 2016). Additionally, SRL is compatible with the modules because it fosters a learning environment that supports students' motivation, enables students to become aware of their own learning strategies, and requires reflection on learning while experiencing the module (Peters and Kitsantas 2010).

The theory behind SRL (see Figure 2.2) explains the different processes that students engage in before, during, and after a learning task. Because SRL is a cyclical learning process, the accomplishment of one cycle develops strategies for the next learning cycle. This cyclic way of learning aligns with the various sections in the STEM Road Map lesson plans on Introductory Activity/Engagement, Activity/Exploration, Explanation, Elaboration/Application of Knowledge, and Evaluation/Assessment. Since the students engaged in a module take on much of the responsibility for learning, this theory also provides guidance for teachers to keep students on the right track.

The remainder of this section explains how SRL theory is embedded within the five sections of each module and points out ways to

Figure 2.2. SRL Theory

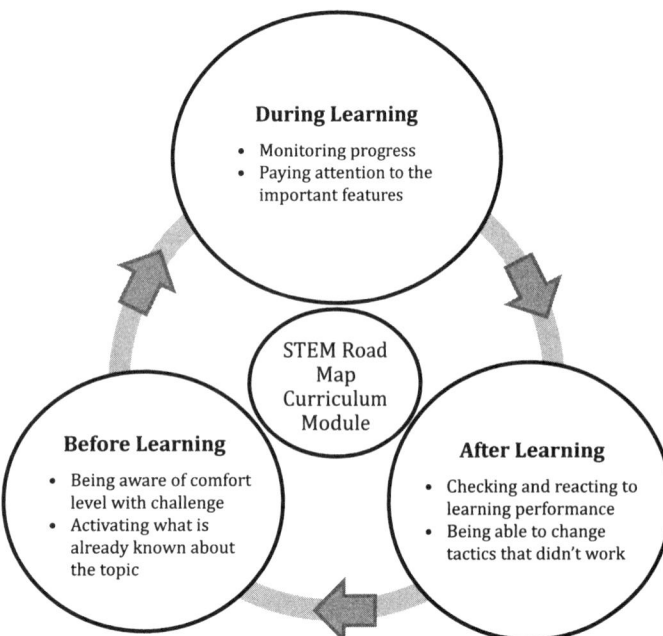

Source: Adapted from Zimmerman 2000.

support students in becoming independent learners of STEM while productively functioning in collaborative teams.

Before Learning: Setting the Stage

Before attempting a learning task such as the STEM Road Map modules, teachers should develop an understanding of their students' level of comfort with the process of accomplishing the learning and determine what they already know about the topic. When students are comfortable with attempting a learning task, they tend to take more risks in learning and as a result achieve deeper learning (Bandura 1986).

The STEM Road Map curriculum modules are designed to foster excitement from the very beginning. Each module has an Introductory Activity/Engagement section that introduces the overall topic from a unique and exciting perspective, engaging the students to learn more so that they can accomplish the challenge. The Introductory Activity also has a design component that helps teachers assess what students already know about the topic of the module. In addition to the deliberate designs in the lesson plans to support SRL, teachers can support a high level of student comfort with the learning challenge by finding out if students have ever accomplished the same kind of task and, if so, asking them to share what worked well for them.

During Learning: Staying the Course

Some students fear inquiry learning because they aren't sure what to do to be successful (Peters 2010). However, the STEM Road Map curriculum modules are embedded with tools to help students pay attention to knowledge and skills that are important for the learning task and to check student understanding along the way. One of the most important processes for learning is the ability for learners to monitor their own progress while performing a learning task (Peters 2012). The modules allow students to monitor their progress with tools such as the STEM Research Notebooks, in which they record what they know and can check whether they have acquired a complete set of knowledge and skills. The STEM Road Map modules support inquiry strategies that include previewing, questioning, predicting, clarifying, observing, discussing, and journaling (Morrison and Milner 2014). Through the use of technology throughout the modules, inquiry is supported by providing students access to resources and data while enabling them to process information, report the findings, collaborate, and develop 21st century skills.

It is important for teachers to encourage students to have an open mind about alternative solutions and procedures (Milner and Sondergeld 2015) when working through the STEM Road Map curriculum modules. Novice learners can have difficulty knowing what to pay attention to and tend to treat each possible avenue for information as equal (Benner 1984). Teachers are the mentors in a classroom and can point out ways for students to approach learning during the Activity/Exploration, Explanation, and

Elaboration/Application of Knowledge portions of the lesson plans to ensure that students pay attention to the important concepts and skills throughout the module. For example, if a student is to demonstrate conceptual awareness of motion when working on roller coaster research, but the student has misconceptions about motion, the teacher can step in and redirect student learning.

After Learning: Knowing What Works

The classroom is a busy place, and it may often seem that there is no time for self-reflection on learning. Although skipping this reflective process may save time in the short term, it reduces the ability to take into account things that worked well and things that didn't so that teaching the module may be improved next time. In the long run, SRL skills are critical for students to become independent learners who can adapt to new situations. By investing the time it takes to teach students SRL skills, teachers can save time later, because students will be able to apply methods and approaches for learning that they have found effective to new situations. In the Evaluation/Assessment portion of the STEM Road Map curriculum modules, as well as in the formative assessments throughout the modules, two processes in the after-learning phase are supported: evaluating one's own performance and accounting for ways to adapt tactics that didn't work well. Students have many opportunities to self-assess in formative assessments, both in groups and individually, using the rubrics provided in the modules.

The designs of the *NGSS* and *CCSS* allow for students to learn in diverse ways, and the STEM Road Map curriculum modules emphasize that students can use a variety of tactics to complete the learning process. For example, students can use STEM Research Notebooks to record what they have learned during the various research activities. Notebook entries might include putting objectives in students' own words, compiling their prior learning on the topic, documenting new learning, providing proof of what they learned, and reflecting on what they felt successful doing and what they felt they still needed to work on. Perhaps students didn't realize that they were supposed to connect what they already knew with what they learned. They could record this and would be prepared in the next learning task to begin connecting prior learning with new learning.

SAFETY IN STEM

Student safety is a primary consideration in all subjects but is an area of particular concern in science, where students may interact with unfamiliar tools and materials that may pose additional safety risks. It is important to implement safety practices within the context of STEM investigations, whether in a classroom laboratory or in the field. When you keep safety in mind as a teacher, you avoid many potential issues with the lesson while also protecting your students.

STEM safety practices encompass things considered in the typical science classroom. Ensure that students are familiar with basic safety considerations, such as wearing protective equipment (e.g., safety glasses or goggles and latex-free gloves) and taking care with sharp objects, and know emergency exit procedures. Teachers should learn beforehand the locations of the safety eyewash, fume hood, fire extinguishers, and emergency shut-off switch in the classroom and how to use them. Also be aware of any school or district safety policies that are in place and apply those that align with the work being conducted in the lesson. It is important to review all safety procedures annually.

STEM investigations should always be supervised. Each lesson in the modules includes teacher guidelines for applicable safety procedures that should be followed. Before each investigation, teachers should go over these safety procedures with the student teams. Some STEM focus areas such as engineering require that students can demonstrate how to properly use equipment in the maker space before the teacher allows them to proceed with the lesson.

Information about classroom science safety, including a safety checklist for science classrooms, general lab safety recommendations, and links to other science safety resources, is available at the Council of State Science Supervisors (CSSS) website at *www.csss-science. org/safety.shtml.* The National Science Teachers Association (NSTA) provides a list of science rules and regulations, including standard operating procedures for lab safety, and a safety acknowledgment form for students and parents or guardians to sign. You can access these resources at *http://static.nsta.org/pdfs/SafetyInTheScienceClassroom.pdf.* In addition, NSTA's Safety in the Science Classroom web page (*www.nsta.org/safety*) has numerous links to safety resources, including papers written by the NSTA Safety Advisory Board.

Disclaimer: The safety precautions for each activity are based on use of the recommended materials and instructions, legal safety standards, and better professional practices. Using alternative materials or procedures for these activities may jeopardize the level of safety and therefore is at the user's own risk.

REFERENCES

Bandura, A. 1986. *Social foundations of thought and action: A social cognitive theory.* Englewood Cliffs, NJ: Prentice-Hall.

Barell, J. 2006. *Problem-based learning: An inquiry approach.* Thousand Oaks, CA: Corwin Press.

Benner, P. 1984. *From novice to expert: Excellence and power in clinical nursing practice.* Menlo Park, CA: Addison-Wesley.

Black, P., C. Harrison, C. Lee, B. Marshall, and D. Wiliam. 2003. *Assessment for learning: Putting it into practice.* Berkshire, UK: Open University Press.

Black, P., and D. Wiliam. 1998. Inside the black box: Raising standards through classroom assessment. *Phi Delta Kappan* 80 (2): 139–148.

Blumenfeld, P., E. Soloway, R. Marx, J. Krajcik, M. Guzdial, and A. Palincsar. 1991. Motivating project-based learning: Sustaining the doing, supporting learning. *Educational Psychologist* 26 (3): 369–398.

Brookhart, S. M., and A. J. Nitko. 2008. *Assessment and grading in classrooms.* Upper Saddle River, NJ: Pearson.

Bybee, R., J. Taylor, A. Gardner, P. Van Scotter, J. Carlson Powell, A. Westbrook, and N. Landes. 2006. *The BSCS 5E instructional model: Origins and effectiveness.* Colorado Springs, CO: BSCS.

Eliason, C. F., and L. T. Jenkins. 2012. *A practical guide to early childhood curriculum.* 9th ed. New York: Merrill.

Johnson, C. 2003. Bioterrorism is real-world science: Inquiry-based simulation mirrors real life. *Science Scope* 27 (3): 19–23.

Krajcik, J., and P. Blumenfeld. 2006. Project-based learning. In *The Cambridge handbook of the learning sciences,* ed. R. Keith Sawyer, 317–334. New York: Cambridge University Press.

Lambros, A. 2004. *Problem-based learning in middle and high school classrooms: A teacher's guide to implementation.* Thousand Oaks, CA: Corwin Press.

Milner, A. R., and T. Sondergeld. 2015. Gifted urban middle school students: The inquiry continuum and the nature of science. *National Journal of Urban Education and Practice* 8 (3): 442–461.

Morrison, V., and A. R. Milner. 2014. Literacy in support of science: A closer look at cross-curricular instructional practice. *Michigan Reading Journal* 46 (2): 42–56.

National Association for the Education of Young Children (NAEYC). 2016. Developmentally appropriate practice position statements. *www.naeyc.org/positionstatements/dap.*

Peters, E. E. 2010. Shifting to a student-centered science classroom: An exploration of teacher and student changes in perceptions and practices. *Journal of Science Teacher Education* 21 (3): 329–349.

Peters, E. E. 2012. Developing content knowledge in students through explicit teaching of the nature of science: Influences of goal setting and self-monitoring. *Science and Education* 21 (6): 881–898.

Peters, E. E., and A. Kitsantas. 2010. The effect of nature of science metacognitive prompts on science students' content and nature of science knowledge, metacognition, and self-regulatory efficacy. *School Science and Mathematics* 110: 382–396.

Popham, W. J. 2013. *Classroom assessment: What teachers need to know.* 7th ed. Upper Saddle River, NJ: Pearson.

Ritchhart, R., M. Church, and K. Morrison. 2011. *Making thinking visible: How to promote engagement, understanding, and independence for all learners.* San Francisco, CA: Jossey-Bass.

Sondergeld, T. A., C. A. Bell, and D. M. Leusner. 2010. Understanding how teachers engage in formative assessment. *Teaching and Learning* 24 (2): 72–86.

Zimmerman, B. J. 2000. Attaining self-regulation: A social-cognitive perspective. In *Handbook of self-regulation,* ed. M. Boekaerts, P. Pintrich, and M. Zeidner, 13–39. San Diego: Academic Press.

PART 2

RAINWATER ANALYSIS

STEM ROAD MAP MODULE

RAINWATER ANALYSIS MODULE OVERVIEW

Paula Schoeff, Janet B. Walton, Carla C. Johnson, and Erin Peters-Burton

THEME: The Represented World

LEAD DISCIPLINE: Mathematics

MODULE SUMMARY

In this module, student teams use the engineering design process (EDP) to plan, construct, and test their own systems for collecting and reusing rainwater to irrigate a fictional community garden. Using their own school building and grounds as a design lab, students learn about measuring rainfall, volume calculations, and rainwater storage options. They also learn about Earth's four spheres and make observations about how the spheres interact. Students explore poetry, biographical texts, and persuasive writing related to module topics. Based on what they have learned, students formulate a message about watershed conservation for their local community and create materials for a public service advertising campaign to disseminate their message (adapted from Capobianco et al. 2015).

ESTABLISHED GOALS AND OBJECTIVES

At the conclusion of this module, students will be able to do the following:

- Create a rain gauge
- Analyze rainwater data to determine the best location for a water collection system
- Use mathematics to explore volume calculations in a variety of real-world scenarios
- Use an electronic spreadsheet to conduct repetitive volume calculations
- Use the EDP to create a design for a rainwater collection and delivery system
- Identify how water is distributed throughout Earth's four spheres

- Research and analyze irrigation and water collection systems used in agriculture

- Present a proposal for a rainwater collection system

- Recognize that the environment can be modified over time by human activities and that these modifications can have both positive and negative consequences

CHALLENGE OR PROBLEM FOR STUDENTS TO SOLVE: THE RAINWATER ROUNDUP CHALLENGE

As the culminating activity for the module, student teams are challenged to design rainwater-recycling systems to provide water for a fictional community garden. Students use information about the interconnectedness of Earth's spheres, mathematical modeling, rainfall analysis, and irrigation as they work toward the goal of designing the systems and creating presentations that describe the design process and the systems' features.

Driving Question: How can we use what we know about rainfall to design a system to provide water to a garden?

CONTENT STANDARDS ADDRESSED IN THIS STEM ROAD MAP MODULE

A full listing with descriptions of the standards this module addresses can be found in the appendix. Listings of the particular standards addressed within lessons are provided in a table for each lesson in Chapter 4.

STEM RESEARCH NOTEBOOK

Each student should maintain a STEM Research Notebook, which will serve as a place for students to organize their work throughout this module (see p. 12 for more general discussion on setup and use of the notebook). All written work in the module should be included in the notebook, including records of students' thoughts and ideas, fictional accounts based on the concepts in the module, and records of student progress through the EDP. The notebooks may be maintained across subject areas, giving students the opportunity to see that although their classes may be separated during the school day, the knowledge they gain is connected.

Each lesson in this module includes student handouts that should be kept in the STEM Research Notebooks after completion, as well as prompts to which students should respond in their notebooks. Students will have the opportunity to create covers and tables of contents for their Research Notebooks in Lesson 1. You may also wish to have students include the STEM Research Notebook Guidelines student handout on page 26 in their notebooks.

Emphasize to students the importance of organizing all information in a Research Notebook. Explain to them that scientists and other researchers maintain detailed Research Notebooks in their work. These notebooks, which are crucial to researchers' work because they contain critical information and track the researchers' progress, are often considered legal documents for scientists who are pursuing patents or wish to provide proof of their discovery process.

STEM RESEARCH NOTEBOOK GUIDELINES

STEM professionals record their ideas, inventions, experiments, questions, observations, and other work details in notebooks so that they can use these notebooks to help them think about their projects and the problems they are trying to solve. You will each keep a STEM Research Notebook during this module that is like the notebooks that STEM professionals use. In this notebook, you will include all your work and notes about ideas you have. The notebook will help you connect your daily work with the big problem or challenge you are working to solve.

It is important that you organize your notebook entries under the following headings:

1. **Chapter Topic or Title of Problem or Challenge:** You will start a new chapter in your STEM Research Notebook for each new module. This heading is the topic or title of the big problem or challenge that your team is working to solve in this module.

2. **Date and Topic of Lesson Activity for the Day:** Each day, you will begin your daily entry by writing the date and the day's lesson topic at the top of a new page. Write the page number both on the page and in the table of contents.

3. **Information Gathered From Research:** This is information you find from outside resources such as websites or books.

4. **Information Gained From Class or Discussions With Team Members:** This information includes any notes you take in class and notes about things your team discusses. You can include drawings of your ideas here, too.

5. **New Data Collected From Investigations:** This includes data gathered from experiments, investigations, and activities in class.

6. **Documents:** These are handouts and other resources you may receive in class that will help you solve your big problem or challenge. Paste or staple these documents in your STEM Research Notebook for safekeeping and easy access later.

7. **Personal Reflections:** Here, you record your own thoughts and ideas on what you are learning.

8. **Lesson Prompts:** These are questions or statements that your teacher assigns you within each lesson to help you solve your big problem or challenge. You will respond to the prompts in your notebook.

9. **Other Items:** This section includes any other items your teacher gives you or other ideas or questions you may have.

MODULE LAUNCH

Launch the module by telling students that they will devise a way to capture and reuse rainwater. Then, hold discussions on the role water plays in supporting life on Earth and the idea that freshwater resources are limited. Show students the slide presentation "Water and Life" at *http://tinyurl.com/WaterAndLife,* which explains the importance of water and its role in supporting life on Earth. Then, ask students to review the distribution of water sources on Earth and discuss the idea that water is a precious resource that must be protected. Next, have the class read together the Rainwater Roundup Challenge scenario to introduce the design challenge for the module. The scenario involves the characters Mamito Anna and Grandpa Henry, who are looking for a solution to the problem of providing water to a garden, as well as Jorge and Angie, students at the fictional Wilbur Wright Elementary School, who help solve the problem. Students learn that Wilbur Wright Elementary resembles their school and grounds in size and shape and are challenged to use their own campus as a design lab to formulate a solution to the problem presented.

PREREQUISITE SKILLS FOR THE MODULE

Students enter this module with a wide range of preexisting skills, information, and knowledge. Table 3.1 (p. 28) provides an overview of prerequisite skills and knowledge that students are expected to apply in this module, along with examples of how they apply this knowledge throughout the module. Differentiation strategies are also provided for students who may need additional support in acquiring or applying this knowledge.

Table 3.1. Prerequisite Key Knowledge and Examples of Applications and Differentiation Strategies

Prerequisite Key Knowledge	Application of Knowledge by Students	Differentiation for Students Needing Additional Knowledge
Science • Water's natural tendency is to flow downhill, and energy is required to move water to a higher level. • Water and air are necessary for life.	*Science* • Identify Earth's water sources and conclude that water circulates through all four of Earth's spheres. • Create models to show that water carries chemicals and pollutants that may affect the environment far away. • Conclude that farmers and gardeners often use irrigation and water collection systems to help provide some of the water needs for fields and gardens.	*Science* • Provide demonstrations and physical models of water movement and irrigation methods. • Review the basic needs of living things. • Provide a variety of nonfiction literature sources about water and the needs of living things.
Inquiry Skills • Ask questions, make logical predictions, plan investigations, and represent data. • Use senses and tools to make observations. • Communicate understanding of data using age-appropriate vocabulary.	*Inquiry Skills* • Select and use appropriate tools and equipment to conduct investigations. • Maintain a notebook that includes observations, data, diagrams, and reflections. • Analyze and communicate findings from multiple investigations of similar phenomena to reach a conclusion.	*Inquiry Skills* • Model selection and use appropriate tools and simple equipment to conduct an investigation. • Provide samples of STEM Research Notebook pages. • Scaffold student efforts to organize data into appropriate tables, graphs, drawings, or diagrams by providing step-by-step instructions. • Identify specific investigations that could be used to answer a particular question, and explain reasons for this choice.

Continued

Table 3.1. (*continued*)

Prerequisite Key Knowledge	Application of Knowledge by Students	Differentiation for Students Needing Additional Knowledge
Numbers and Representation • Convert numbers from fractions to decimals. • Understand place value of decimal numbers. • Multiply, add, and subtract decimal numbers to thousandths place. • Understand that measurements expressed as numbers have various units associated with them.	*Numbers and Representation* • Use mathematics operations when creating tables and graphs for storing and analyzing data. • Express measurements using appropriate units.	*Numbers and Representation* • Review properties of operations using examples of volume and distance. • Use textbook support, teacher instruction, models, graphic organizers, and online instruction to provide practice.
Measurement Skills • Measure volume and distance with appropriate tools and units. • Make precise measurements. • Convert units of measurement. • Understand the concept of scale, and create scaled drawings. • Understand how to use a protractor.	*Measurement Skills* • Make linear measurements and convert them to their decimal equivalents. • Measure volumes of water using cylinders and measuring cups. • Construct their own measurement tools to solve specific problems. • Convert units of measurement. • Create scaled drawings of the school grounds and school building. • Use a protractor to create a tool to measure angles in the schoolyard.	*Measurement Skills* • Provide opportunities to practice measuring with precision, using the correct units. • Provide a table that offers visual reinforcement of measurements and units. • Provide instruction in use of cylinders and measuring cups for finding volume. • Provide handouts to offer guidance in measuring and conversion. • Provide examples of scale using maps, and model the use of scale by working as a class to create a scaled drawing of the classroom. • Review protractor use with the class, providing opportunities for students to use protractors to measure both inside and outside angles in the classroom.

Continued

Table 3.1. (*continued*)

Prerequisite Key Knowledge	Application of Knowledge by Students	Differentiation for Students Needing Additional Knowledge
Geometry • Identify quadrants on a coordinate grid. • Identify points on a coordinate grid. • Recognize three-dimensional (3-D) geometric shapes. • Understand that modeling scales are a ratio of the measurement on a drawing compared with the measurement of the original object. • Understand and measure angles and understand the principle of opposite angles.	*Geometry* • Plot points on a coordinate grid. • Identify and employ fundamental geometric shapes in real-world structures. • Make multiview drawings of 3-D objects. • Measure angles in the schoolyard, using the principle of opposite angles to measure exterior angles.	*Geometry* • Demonstrate the use of coordinate grids, and provide opportunities for students to practice plotting points and identifying quadrants. • Provide opportunities to measure 3-D cubes and rectangular solids. • Provide opportunities to find the volume of cylindrical shapes. • Demonstrate the principle of opposite angles, and provide opportunities to apply this concept to measure exterior angles in the classroom.
Reading Skills • Read grade-level science texts, and decode words using phonics and word analysis skills. • Use information gained from illustrations and text to understand science concepts. • Draw inferences from informational text.	*Reading Skills* • Research and report on the importance of water on Earth. • Use the internet and grade-appropriate texts to conduct research. • Read fiction and nonfiction texts that portray topics from the module in a variety of ways.	*Reading Skills* • Provide reading strategies to support comprehension of nonfiction texts, including using vocabulary notecards and games, graphic organizers, STEM Research Notebooks, and discussions. • Through class read-alouds, model interpreting illustrations and graphics in texts and drawing inferences from informational texts.
Writing Skills • Use science terms to write informative texts and explain thoughts and ideas about rainwater collection systems. • Use key terminology as words and pictures. • Provide evidence to support ideas and opinions about topics.	*Writing Skills* • Write informative and explanatory narratives to convey ideas and information clearly. • Write narratives to describe experiences using effective techniques, descriptive details, and clear event sequences.	*Writing Skills* • Provide templates or graphic organizers for writing. • Model organizational techniques for writing. • Provide rubrics for students to assess their own writing.

Continued

Table 3.1. (*continued*)

Prerequisite Key Knowledge	Application of Knowledge by Students	Differentiation for Students Needing Additional Knowledge
Communication Skills • Participate in collaborative conversations using appropriate language and skills. • Effectively support scientific knowledge with appropriate language and relevant, descriptive details.	*Communication Skills* • Engage in a number of collaborative discussions that support learning. • Create a presentation to share classroom experiences and research.	*Communication Skills* • Scaffold student understanding of communication skills by providing examples of appropriate language and presentation. • Provide handouts and rubrics to support organization of facts and use of relevant descriptive details.

POTENTIAL STEM MISCONCEPTIONS

Students enter the classroom with a wide variety of prior knowledge and ideas, so it is important to be alert to misconceptions, or inappropriate understandings of foundational knowledge. These misconceptions can be classified as one of several types: "preconceived notions," opinions based on popular beliefs or understandings; "nonscientific beliefs," knowledge students have gained about science from sources outside the scientific community; "conceptual misunderstandings," incorrect conceptual models based on incomplete understanding of concepts; "vernacular misconceptions," misunderstandings of words based on their common use versus their scientific use; and "factual misconceptions," incorrect or imprecise knowledge learned in early life that remains unchallenged (NRC 1997, p. 28). Misconceptions must be addressed and dismantled in order for students to reconstruct their knowledge, and therefore teachers should be prepared to take the following steps:

- *Identify students' misconceptions.*

- *Provide a forum for students to confront their misconceptions.*

- *Help students reconstruct and internalize their knowledge, based on scientific models.* (NRC 1997, p. 29)

Keeley and Harrington (2010) recommend using diagnostic tools such as probes and formative assessment to identify and confront student misconceptions and begin the process of reconstructing student knowledge. Keeley's *Uncovering Student Ideas in Science* series contains probes targeted toward uncovering student misconceptions in a variety of areas and may be a useful resource for addressing student misconceptions in this module.

Some commonly held misconceptions specific to lesson content are provided with each lesson so that you can be alert for student misunderstanding of the science concepts presented and used during this module. The American Association for the Advancement

of Science has also identified misconceptions that students frequently hold regarding various science concepts (see the links at *http://assessment.aaas.org/topics*).

SRL PROCESS COMPONENTS

Table 3.2 illustrates some of the activities in the Rainwater Analysis module and how they align to the self-regulated learning (SRL) process before, during, and after learning.

Table 3.2. SRL Process Components

Learning Process Components	Example From Rainwater Analysis Module	Lesson Number and Learning Component
BEFORE LEARNING		
Motivates students	Students are told that they will devise a way to capture and reuse rainwater. Then, they discuss the role water plays in supporting life on Earth and the idea that freshwater resources are limited.	Lesson 1, Introductory Activity/Engagement
Evokes prior learning	Students use prior knowledge in discussing water's role in supporting life on Earth.	Lesson 1, Introductory Activity/Engagement
DURING LEARNING		
Focuses on important features	Students complete a home survey of features that handle rainwater, then share their findings with classmates.	Lesson 2, Activity/Exploration
Helps students monitor their progress	Student teams compare their rain gauge data with the other teams' data and with rainfall data from their area. Students discuss reasons for variations in data.	Lesson 2, Elaboration/Application of Knowledge
AFTER LEARNING		
Evaluates learning	Students get feedback on their final challenge product from detailed rubrics.	Lesson 4, Elaboration/Application of Knowledge
Takes account of what worked and what did not work	Students reflect on the feedback they receive when they present to the principal and other stakeholders.	Lesson 4, Elaboration/Application of Knowledge

STRATEGIES FOR DIFFERENTIATING INSTRUCTION WITHIN THIS MODULE

For the purposes of this curriculum module, differentiated instruction is conceptualized as a way to tailor instruction—including process, content, and product—to various student needs in your class. A number of differentiation strategies are integrated into lessons across the module. The problem- and project-based learning approach used in the lessons is designed to address students' multiple intelligences by providing a variety of entry points and methods to investigate the key concepts in the module (for example, investigating rainwater and horticulture using scientific inquiry, fiction and nonfiction literature, journaling, and collaborative design). Differentiation strategies for students needing support in prerequisite knowledge can be found in Table 3.1 (p. 28). You are encouraged to use information gained about student prior knowledge during introductory activities and discussions to inform your instructional differentiation. Strategies incorporated into this lesson include flexible grouping, varied environmental learning contexts, assessments, compacting, tiered assignments and scaffolding, and mentoring.

Flexible Grouping. Students work collaboratively in a variety of activities throughout this module. Grouping strategies you may choose to employ include student-led grouping, grouping students according to ability level or common interests, grouping students randomly, or grouping them so that students in each group have complementary strengths (for instance, one student might be strong in mathematics, another in art, and another in writing).

Varied Environmental Learning Contexts. Students have the opportunity to learn in various contexts throughout the module, including alone, in groups, in quiet reading and research-oriented activities, and in active learning through inquiry and design activities. In addition, students learn in a variety of ways, including through doing inquiry activities, reading a variety of texts, writing about module topics using various genres of literature, watching videos, participating in class discussion, and conducting web-based research.

Assessments. Students are assessed in a variety of ways throughout the module, including individual and collaborative formative and summative assessments. Students have the opportunity to produce work via written text, oral and media presentations, and modeling. You may choose to provide students with additional choices of media for their products (for example, electronic slide presentations, posters, or student-created websites or blogs).

Compacting. Based on student prior knowledge, you may wish to adjust instructional activities for students who exhibit prior mastery of a learning objective. Because student work in science and mathematics is largely collaborative throughout the module, this strategy may be most appropriate for ELA or social studies activities. You may wish to compile a classroom database of research resources and supplementary readings for a variety of reading levels and on a variety of topics related to the module's topic to provide opportunities for students to undertake independent reading.

Tiered Assignments and Scaffolding. Based on your awareness of student ability, understanding of concepts, and mastery of skills, you may wish to provide students with variations on activities by adding complexity to assignments or providing more or fewer learning supports for activities throughout the module. For instance, some students may need additional support in identifying key search words and phrases for web-based research or may benefit from cloze sentence handouts to enhance vocabulary understanding. Other students may benefit from expanded reading selections and additional reflective writing or from working with manipulatives and other visual representations of mathematical concepts. You may also work with your school librarian to compile a set of topical resources at a variety of reading levels.

Mentoring. As group design teamwork becomes increasingly complex throughout the module, you may wish to have a resource teacher, older student, or parent volunteer work with groups that struggle to stay on task and collaborate effectively.

STRATEGIES FOR ENGLISH LANGUAGE LEARNERS

Students who are developing proficiency in English language skills require additional supports to simultaneously learn academic content and the specialized language associated with specific content areas. WIDA (2012) has created a framework for providing support to these students and makes available rubrics and guidance on differentiating instructional materials for English language learners (ELLs). In particular, ELL students may benefit from additional sensory supports such as images, physical modeling, and graphic representations of module content, as well as interactive support through collaborative work. This module incorporates a variety of sensory supports and offers ongoing opportunities for ELL students to work collaboratively. The focus on rainwater affords an opportunity for ELL students to share culturally diverse experiences with climate conditions, horticulture, and agriculture.

Teachers differentiating instruction for ELL students should carefully consider the needs of these students as they introduce and use academic language in various language domains (listening, speaking, reading, and writing) throughout this module. To adequately differentiate instruction for ELL students, teachers should have an understanding of the proficiency level of each student. The following five preK–5 WIDA learning standards are relevant to this module:

- Standard 1: Social and Instructional Language. Report on a topic or text, tell a story, or recount an experience in an organized manner, using appropriate facts and relevant, descriptive details to support main ideas or themes; speak clearly at an understandable pace; add audio recordings and visual displays to presentations when appropriate to enhance the development of main ideas or themes.

- Standard 2: The language of Language Arts. Quote accurately from a text when explaining what the text says explicitly and when drawing inferences from the text.

- Standard 3: The language of Mathematics. Use a pair of perpendicular number lines, called axes, to define a coordinate system. Represent real world and mathematical problems by graphing points in the first quadrant of the coordinate plane and interpret coordinate values of points in the context of the situation.

- Standard 4: The language of Science. Focus on forces in nature, scientific process, Earth and sky, living and nonliving things, organisms and environment, and weather.

- Standard 5: The language of Social Studies. Integrate visual information (e.g., in charts, graphs, photographs, videos, or maps) with other information in print and digital texts.

SAFETY CONSIDERATIONS FOR THE ACTIVITIES IN THIS MODULE

This module's science component focuses on water and its interaction with living things. Ensure that any water spilled on the floor is cleaned up promptly to avoid slipping. See the safety notes in each lesson pertaining to the specific activities in that lesson. For more general safety guidelines, see the Safety in STEM section in Chapter 2 (p. 18).

DESIRED OUTCOMES AND MONITORING SUCCESS

The desired outcomes for this module are outlined in Table 3.3 (p. 36), along with suggested ways to gather evidence to monitor student success. For more specific details on desired outcomes, see the Established Goals and Objectives sections for the module and individual lessons.

Table 3.3. Desired Outcomes and Evidence of Success in Achieving Identified Outcomes

Desired Outcomes	Evidence of Success	
	Performance Tasks	Other Measures
• Students recognize water's essential role to life on Earth. • Students can apply an understanding of the hydrosphere and water collection to create a model of a rainwater collection system. • Students share their learning through writing and presentations.	• Students maintain STEM Research Notebooks that contain graphic organizers with data from investigations, sketches, research notes, evidence of collaboration, and ELA-related work. • Students use their school playground as a design lab to plan a rainwater-recycling system to be used in a fictional garden presented in a scenario. • Students can defend their design decisions. • Students are assessed using rubrics that focus on learning and application of skills related to the academic content.	• STEM Research Notebooks are assessed using a rubric. • Student collaboration is evaluated using self-assessment reflections, peer feedback, and a collaboration rubric.

ASSESSMENT PLAN OVERVIEW AND MAP

Table 3.4 provides an overview of the major group and individual *products* and *deliverables*, or things that student teams will produce in this module, that constitute the assessment for this module. See Table 3.5 (p. 38) for a full assessment map of formative and summative assessments in this module.

Table 3.4. Major Products and Deliverables in Lead Disciplines for Groups and Individuals

Lesson	Major Group Products and Deliverables	Major Individual Products and Deliverables
1	• Playing Card Challenge • Rain Gauge Design Challenge • Watershed conservation research	• Watershed Model • Watershed Place Mat • Volume Conversion Table • Original poem • STEM Research Notebook entries
2	• Surveyor Tools • Schoolyard Surveyors • Schoolyard Map • Build Your Own Biodome	• Rainwater at Home Survey • Earth's Spheres Poster • Biography of a Classmate • STEM Research Notebook entries
3	• What's the Volume? • How Big Is Big Enough? • Writing the OREO Way as a group • Irrigation research • Way to Flow: Capillary Action • The Great Escape: Capillary Action	• Writing the OREO Way individually • Understanding Weather Data • STEM Research Notebook entries
4	• Cylinder Volume spreadsheet • Collection Tank Design Challenge • Garden Water Investigation • Water Distribution Design Challenge • Rainwater Roundup Challenge • Rainwater Roundup Slideshow • Public Service Advertising Campaign materials	• STEM Research Notebook entries

Table 3.5. Assessment Map for Rainwater Analysis Module

Lesson	Assessment	Group/ Individual	Formative/ Summative	Lesson Objective Assessed
1	Playing Card Challenge *handout*	Group	Formative	• Calculate the volume of a small item that cannot be physically measured with the tools available.
1	Rain Gauge Design Challenge *rubric*	Group	Formative	• Design a water gauge and place several throughout the schoolyard. • Track the amount of water that falls in a variety of locations throughout the schoolyard.
1	Watershed Model *activity*	Individual	Formative	• Recognize the value of water on Earth.
1	Watershed research and Watershed Place Mat *rubric*	Individual	Formative	• Research the importance of a community's watershed and propose ways that farmers and the community might protect it.
1	Volume Conversion Table	Individual	Formative	• Calculate the volume of a small item that cannot be physically measured with the tools available.
1	Poetry Writing *rubric*	Individual	Formative	• Understand that poetry can be used to express feelings and communicate ideas about phenomena in the natural world. • Use this understanding to create an original poem related to module topics.
2	Rainwater at Home Survey *handout*	Individual	Formative	• Identify features of buildings that keep rainwater from entering the buildings.
2	Surveyor Tools *handout*	Group	Formative	• Understand that specialized instruments are used to measure large spaces. • Design measuring instruments to measure the school building and schoolyard.

Continued

NATIONAL SCIENCE TEACHERS ASSOCIATION

Table 3.5. (*continued*)

Lesson	Assessment	Group/ Individual	Formative/ Summative	Lesson Objective Assessed
2	Schoolyard Surveyors *handout*	Group	Formative	• Use measuring instruments to measure the school building and schoolyard. • Recognize features of buildings that serve to keep rainwater outside the buildings. • Use observations of rainwater handling features to predict where rainwater goes when it is channeled away from the building. • Measure the footprint (amount of space covered by an object) of large structures such as the school building and grounds.
2	Schoolyard Map *rubric*	Group	Formative	• Build a scale map of the schoolyard.
2	Build Your Own Biodome *handout* or EDP Applied to the Biodome *handout*	Group	Formative	• Create a biodome model, a small terrarium that mimics the conditions of natural ecosystems.
2	Map Detective *activity*	Group	Formative	• Use maps to make inferences about how Earth's major systems interact.
2	Earth's Spheres *poster*	Individual	Formative	• Identify the four Earth spheres (biosphere, geosphere, hydrosphere, and atmosphere). • Provide examples of the ways that each of Earth's four systems interact with each other. • Create a model showing how Earth's major spheres interact.
2	Biography of a Classmate *rubric*	Individual	Formative	• Identify the characteristics of biographical writing. • Identify the difference between primary and secondary sources. • Write a biographical text.
2	Collaboration *rubric*	Individual	Formative	• Use collaboration skills to accomplish work as members of a team.

Continued

Table 3.5. (*continued*)

Lesson	Assessment	Group/ Individual	Formative/ Summative	Lesson Objective Assessed
3	What's the Volume? *handout*	Group	Formative	• Calculate the volume of rainwater that falls on the playground on one morning.
3	How Big Is Big Enough? *handout*	Group	Formative	• Design a tank to hold a given amount of water.
3	Way to Flow: Capillary Action *handout*	Group	Formative	• Describe capillary action and identify it as the mechanism by which water is moved from soil to various parts of plants.
3	Writing the OREO Way *handout*	Individual/ Group	Formative	• Propose some ways that humans can protect the hydrosphere. • Construct a message about watershed protection using persuasive writing techniques.
3	Irrigation *research*	Group	Formative	• Identify and describe various irrigation techniques. • Evaluate agricultural practices and propose alternative strategies for those that may have harmful impacts on the environment.
3	Understanding Weather Data *spreadsheet or table and graph*	Individual	Formative	• Collect and graph historical rainfall data. • Create and analyze a graph to compare and contrast schoolyard rainfall measurements with statistics for rainfall in the surrounding area in a typical year.
3	The Great Escape: Capillary Action *handout*	Group	Formative	• Provide examples of ways that human activities (the biosphere) can negatively impact Earth's other spheres.
4	Cylinder Volume *spreadsheet tables and graphs*	Group	Formative	• Use a spreadsheet to create a way to conduct repetitive calculations. • Calculate the volume of rectangular solids and cylinders of given dimensions.

Continued

Table 3.5. (*continued*)

Lesson	Assessment	Group/ Individual	Formative/ Summative	Lesson Objective Assessed
4	Collection Tank Design Challenge *rubric*	Group	Summative	• Design a system to store a specified amount of water.
4	Garden Water Investigation *handout*	Group	Summative	• Calculate the volume of rectangular solids and cylinders of given dimensions. • Use understanding of irrigation techniques to propose an irrigation system for a garden.
4	Water Distribution Design Challenge *rubric*	Group	Summative	• Create a model of an irrigation system.
4	Rainwater Roundup Challenge *rubric*	Group	Summative	• Synthesize learning from the module to create a model of a rainwater capture system that includes rainwater storage and a way to deliver stored rainwater to a garden.
4	Rainwater Roundup Slideshow *rubric*	Group	Summative	• Create a slideshow presentation for the rainwater capture system and the process used to create the system.
4	Public Service Advertising Campaign *rubric*	Group	Summative	• Synthesize learning from the module to create persuasive text to convey a message about watershed protection to the public in a media format.

MODULE TIMELINE

Tables 3.6–3.10 (pp. 42–46) provide lesson timelines for each week of the module. The timelines are provided for general guidance only and are based on class times of approximately 45 minutes.

Table 3.6. STEM Road Map Module Schedule for Week One

Day 1	Day 2	Day 3	Day 4	Day 5
Lesson 1 *Water, Water Everywhere!*	*Lesson 1* *Water, Water Everywhere!*	*Lesson 1* *Water, Water Everywhere!*	*Lesson 1* *Water, Water Everywhere!*	*Lesson 1* *Water, Water Everywhere!*
• Launch the module with discussions on the role water plays in supporting life on Earth and the idea that freshwater resources are limited.	• Use a demonstration and an electronic spreadsheet to illustrate the distribution of potable water on Earth.	• Students construct conversion tables.	• Students present and discuss their conversion tables.	• Students begin to devise a data collection plan for their rain gauges.
• View a slideshow that provides an overview of the water cycle and potable water and its importance for life on Earth.	• Introduce concepts associated with measuring rainfall and measuring volume.	• Students begin the Playing Card Challenge, in which they calculate the volume of a single playing card.	• Students complete the Playing Card Challenge.	• Students begin to create their own poetry related to the module topic.
• Introduce the Rainwater Roundup Challenge, using a scenario about watering needs for a garden at the Sunny Acres retirement home.	• Conduct a demonstration about volume.	• Introduce the Rain Gauge Design Challenge.	• Students build rain gauges in the Rain Gauge Design Challenge and place them in locations around the school grounds.	• Students conduct research about the watershed and watershed protection.
	• Introduce Volume Conversion Table activity.	• Introduce several types of poetry.	• Continue to discuss and provide examples (related to the module topic) of types of poetry.	
	• Introduce poetry related to rain.	• Students create a watershed model.	• Hold a class discussion about the importance of the watershed and potential effects of human activities on the watershed.	
	• Introduce the concept of the watershed.			

Table 3.7. STEM Road Map Module Schedule for Week Two

Day 6	Day 7	Day 8	Day 9	Day 10
Lesson 1 *Water, Water Everywhere!* • Students practice calculating volume for rectangular solids using an example of calculating the volume of paint needed to paint their classroom. • The class finalizes the collection plan for rain data. • Students continue to write their own poetry based on the module topic. • Students create place mats that synthesize their research about the watershed and watershed protection.	*Lesson 1* *Water, Water Everywhere!* • Introduce the concept of engineering design by a read-aloud of a book about the design of the Mars rovers. • Students share their poetry and provide feedback to others about their poems. • Students share their place mats with the class.	*Lesson 2* *Earth's Spheres* • Introduce and discuss the idea that Earth is a system made up of subsystems known as spheres. • Begin the Earth's Spheres Poster activity. • Read aloud *Rachel Carson and Her Book That Changed the World* by Laurie Lawlor as an example of biographical writing and how writing can inform the public about environmental issues. • Discuss science and engineering careers and their relationship to Earth's spheres.	*Lesson 2* *Earth's Spheres* • Begin the Rainwater at Home activity. • Introduce concepts associated with measuring large areas. • Begin the Schoolyard Surveyors activity. • Continue the investigation of Earth's spheres with a discussion about how the spheres interact with plant growth. • Introduce the Build Your Own Biodome activity. • Students investigate biographies. • Introduce the idea that geography influences how Earth's spheres appear and how they interact. • Begin the Map Detective activity.	*Lesson 2* *Earth's Spheres* • Complete the Rainwater at Home activity. • Complete the Schoolyard Surveyors activity. • Students create biodomes in the Build Your Own Biodome activity. • Introduce the idea of information sources for biographies. • Students formulate interview questions to ask classmates. • Continue the Map Detective activity.

Table 3.8. STEM Road Map Module Schedule for Week Three

Day 11	Day 12	Day 13	Day 14	Day 15
Lesson 2 *Earth's Spheres* • Begin the Schoolyard Map activity. • Complete biodomes and hold a class discussion about the features of the biodome. • Students interview classmates and share findings. • Complete the Map Detective activity.	*Lesson 2* *Earth's Spheres* • Complete the Schoolyard Map activity. • Students calculate the volume of rain that falls on the school building during a rain event. • Students begin to write biographies about their classmates based on their interview source information.	*Lesson 2* *Earth's Spheres* • Students brainstorm ways to collect rainwater from the school building. • Students share and discuss their solutions for calculating rain volume. • Students continue to write their biographies.	*Lesson 2* *Earth's Spheres* • Students share their rainwater collection ideas and consider the associated challenges. • Students practice rainwater volume calculations using various scenarios. • Students share their biographies and provide and receive feedback.	*Lesson 3* *How Much Rain Can We Catch?* • Introduce the idea of hydrosphere protection in the community using a slideshow and class discussion. • Discuss water shortages and their implications for communities and for agriculture. • Introduce public service advertising (PSA) and persuasive text.

Table 3.9. STEM Road Map Module Schedule for Week Four

Day 16	Day 17	Day 18	Day 19	Day 20
Lesson 3 *How Much Rain Can We Catch?*	*Lesson 3* *How Much Rain Can We Catch?*	*Lesson 3* *How Much Rain Can We Catch?*	*Lesson 3* *How Much Rain Can We Catch?*	*Lesson 4* *The Rainwater Roundup Challenge*
• Optional: Have the school custodian or building manager visit the class to review students' schoolyard maps and provide feedback and additional information. • Begin What's the Volume? activity to calculate rainfall on the school playground. • Introduce the concept of capillary action. • Do the Way to Flow: Capillary Action activity. • Students explore persuasive writing and begin to formulate a message about watershed protection for their community.	• Complete the What's the Volume? activity. • Calculate rainwater tank collection volumes in the How Big Is Big Enough? activity. • Continue to create and refine a message about watershed protection for the community. • Discuss the governmental decision-making processes for water usage. • Introduce irrigation research.	• Introduce the Understanding Weather Data activity. • Do the Great Escape: Capillary Action activity. • Students consider ways to disseminate their information about watershed protection in a PSA campaign. • Continue irrigation research.	• Complete Understanding Weather Data activity. • Students compare their rain gauge data with overall rainfall statistics for their area. • Students agree on a class message about community watershed protection and decide on the media each team will be responsible for. • Optional: Students take a farm or greenhouse field trip or listen to an in-class speaker.	• Review how the EDP will be applied to the module challenge. • Do the Acme Tank Works activity. • Student teams work on creating their media components for the class PSA campaign about watershed protection in their community.

Table 3.10. STEM Road Map Module Schedule for Week Five

Day 21	Day 22	Day 23	Day 24	Day 25
Lesson 4 *The Rainwater Roundup Challenge* • Complete the Collection Tank Design Challenge. • Student teams work on creating their media components for the class PSA campaign.	*Lesson 4* *The Rainwater Roundup Challenge* • Complete the Garden Water Investigation. • Student teams continue to work on their media components for the class PSA campaign.	*Lesson 4* *The Rainwater Roundup Challenge* • Complete the Water Distribution Design Challenge. • Students begin creating their slideshow presentations. • Student teams continue to work on their media components for the class PSA campaign.	*Lesson 4* *The Rainwater Roundup Challenge* • Complete the Rainwater Roundup Challenge. • Students complete their slideshow presentations. • Student teams complete their media components for the class PSA campaign.	*Lesson 4* *The Rainwater Roundup Challenge* • Students give presentations of their Rainwater Roundup slideshows and media components of the class PSA campaign.

RESOURCES

The media specialist can help teachers locate resources for students to view and read about recreational equipment, parks, and related physics content. Special educators and reading specialists can help find supplemental sources for students needing extra support in reading and writing. Additional resources may be found online. Community resources for this module may include civil engineers, environmental engineers, and rainwater handling product manufacturing representatives.

REFERENCES

Capobianco, B. M., C. Parker, A. Laurier, and J. Rankin. 2015. The STEM Road Map for grades 3–5. In *STEM Road Map: A framework for integrated STEM education,* ed. C. C. Johnson, E. E. Peters-Burton, and T. J. Moore, 68–95. New York: Routledge. *www.routledge.com/products/9781138804234.*

Keeley, P., and R. Harrington. 2010. *Uncovering student ideas in physical science, volume 1: 45 new force and motion assessment probes.* Arlington, VA: NSTA Press.

National Research Council (NRC). 1997. *Science teaching reconsidered: A handbook.* Washington, DC: National Academies Press.

WIDA. 2012. 2012 amplification of the English language development standards: Kindergarten–grade 12. *https://wida.wisc.edu/teach/standards/eld.*

RAINWATER ANALYSIS LESSON PLANS

Paula Schoeff, Janet B. Walton, Carla C. Johnson, and Erin Peters-Burton

Lesson Plan 1: Water, Water Everywhere!

This lesson introduces students to the module challenge, the Rainwater Roundup Challenge. Through a slideshow and water activity, students learn about water as a scarce natural resource. As a foundation for understanding the relationship between standard rainfall measurements and volume, students are challenged to determine the volume of a single playing card. Students design rain gauges and place them around the school campus to obtain data for their decisions about where to locate their rain-water collection systems. Students also investigate poetry and write their own poems related to module topics.

ESSENTIAL QUESTIONS

- How can we measure the volume of a very thin item, such as a single card in a deck?

- How can we estimate the amount of rain that falls in a given area?

- What resources can be used to find out how much rain falls in our town in a given year?

- What is meant by the statement "Water is the most essential element for life, and the future of humanity depends on our capacity to guard it and share it"?

- What factors should be considered when making a rain gauge?

- What changes might help protect the watershed in an agricultural community?

ESTABLISHED GOALS AND OBJECTIVES

At the conclusion of this lesson, students will be able to do the following:

- Visualize the amount of potable water available on Earth

- Understand that water is essential for life

- Create a graph to illustrate the distribution of water in the hydrosphere

- Calculate the volume of a small item that cannot be physically measured with the tools available

- Design and build a water gauge

- Track the amount of water that falls in a variety of locations throughout the schoolyard

- Research the importance of a community's watershed and propose ways that farmers and the community might protect it

- Understand that poetry can be used to express feelings and communicate ideas about phenomena in the natural world, and use this understanding to create an original poem related to module topics

TIME REQUIRED

- 7 days (approximately 45 minutes each; see Tables 3.6 and 3.7, pp. 42–43)

MATERIALS

Necessary Materials for Lesson 1

- STEM Research Notebooks (1 per student; see p. 26 for the STEM Research Notebook Guidelines handout)

- Internet access for slideshow and student research

- Handouts and rubrics (attached at the end of this lesson)

Additional Materials for Introductory Activity/Engagement

- Beaker (500 ml) or graduated measuring cup (2 cups)

- Eyedropper

- 1 liter of colored water in a pitcher (made by mixing food coloring or a fruit drink powder into plain water)

- 5 clear plastic cups

- Permanent marker

- Salt

- Indirectly vented chemical splash goggles (1 per student)

- Nonlatex apron (1 per student)
- Rainwater Roundup handout (1 per student; see p. 84)

Additional Materials for Mathematics Class

- Several flashlights
- Coffee mug that is opaque
- Pitcher of water
- Paperback book
- Math cube for 1 cm (or 1 cm game die)
- Team materials (per team unless otherwise noted)
  - Deck of cards
  - Pitcher of water
  - Set of measuring cups
  - Graduated cylinder, 1,000 ml
  - Beaker, 1,000 ml
  - Ruler (with metric and standard units)
  - Calculator
  - Indirectly vented chemical splash goggles (1 per student)
  - Nonlatex apron (1 per student)
  - Volume Conversion Table handout (p. 87)
  - Playing Card Challenge handout (p. 89)

Additional Materials for Science Connection (per team unless otherwise noted)

- Small pitcher
- Large watering can
- Plastic jar or container with vertical sides (e.g., peanut butter container)
- Duct tape
- Vinyl or plastic sign material
- Laminating sheets or resealable plastic bags

- Picture book about Mars rovers such as *Curiosity: The Story of a Mars Rover,* by Markus Motum (Candlewick Press, 2018; 1 per class)

- Indirectly vented chemical splash goggles (1 per student)

- Nonlatex apron (1 per student)

- Rain Gauge EDP (p. 90)

Additional Materials for Social Studies Connection (per student)

- 8 ½ × 11 inch sheet of paper

- 4 washable markers: brown, blue, red, and green

- Misting spray bottle (1 for every 4 students)

- 11 × 17 inch piece of white construction paper

Additional Materials for ELA Connection

- *National Geographic Book of Nature Poetry: More Than 200 Poems With Photographs That Float, Zoom, and Bloom!* by J. Patrick Lewis (National Geographic Children's Books, 2015)

- Picture books about water topics and agriculture (see suggested books list on p. 82)

SAFETY NOTES

1. Direct teacher supervision is imperative during all aspects of this activity to make sure students follow the safety guidelines.

2. All laboratory occupants must wear indirectly vented chemical splash goggles and aprons during all phases of this inquiry activity.

3. Immediately wipe up any water that is spilled on the floor to avoid a slip-and-fall hazard.

4. Use caution when working with sharps (e.g., wire) to avoid cutting or puncturing skin.

5. Handle glassware and plasticware with care, as these can break and cut or puncture skin.

6. Make sure all materials are put away after completing the activity.

7. Wash hands with soap and water after completing the activity.

CONTENT STANDARDS AND KEY VOCABULARY

Table 4.1 lists the content standards from the *Next Generation Science Standards* (*NGSS*), *Common Core State Standards* (*CCSS*), and the Framework for 21st Century Learning that this lesson addresses, and Table 4.2 (p. 59) presents the key vocabulary. Vocabulary terms are provided for both teacher and student use. Teachers may choose to introduce some or all of the terms to students.

Table 4.1. Content Standards Addressed in STEM Road Map Module Lesson 1

NEXT GENERATION SCIENCE STANDARDS

PERFORMANCE EXPECTATIONS

- 5-ESS2-1. Develop a model using an example to describe ways in which the geosphere, biosphere, hydrosphere, and/or atmosphere interact.

- 5-ESS2-2. Describe and graph the amounts and percentages of water and fresh water in various reservoirs to provide evidence about the distribution of water on Earth.

- 5-ESS3-1. Obtain and combine information about ways individual communities use science ideas to protect the Earth's resources and environment.

SCIENCE AND ENGINEERING PRACTICES

Asking Questions and Defining Problems

- Ask questions that can be investigated and predict reasonable outcomes based on patterns such as cause and effect relationships.

Developing and Using Models

- Identify limitations of models.

- Collaboratively develop and/or revise a model based on evidence that shows the relationships among variables for frequent and regular occurring events.

- Develop a model using an analogy, example, or abstract representation to describe a scientific principle or design solution.

- Develop and/or use models to describe and/or predict phenomena.

- Use a model to test cause and effect relationships or interactions concerning the functioning of a natural or designed system.

Planning and Carrying Out Investigations

- Plan and conduct an investigation collaboratively to produce data to serve as the basis for evidence, using fair tests in which variables are controlled and the number of trials considered.

- Evaluate appropriate methods and/or tools for collecting data.

- Make observations and/or measurements to produce data to serve as the basis for evidence for an explanation of a phenomenon or test a design solution.

- Make predictions about what would happen if a variable changes.

Continued

Table 4.1. (*continued*)

Analyzing and Interpreting Data

- Represent data in tables and/or various graphical displays (bar graphs, pictographs, and/or pie charts) to reveal patterns that indicate relationships.

- Analyze and interpret data to make sense of phenomena, using logical reasoning, mathematics, and/or computation.

- Compare and contrast data collected by different groups in order to discuss similarities and differences in their findings.

Using Mathematics and Computational Thinking

- Organize simple data sets to reveal patterns that suggest relationships.

- Describe, measure, estimate, and/or graph quantities (e.g., area, volume, weight, time) to address scientific and engineering questions and problems.

- Create and/or use graphs and/or charts generated from simple algorithms to compare alternative solutions to an engineering problem.

Constructing Explanations and Designing Solutions

- Construct an explanation of observed relationships (e.g., the distribution of plants in the backyard).

- Use evidence (e.g., measurements, observations, patterns) to construct or support an explanation or design a solution to a problem.

- Identify the evidence that supports particular points in an explanation.

- Apply scientific ideas to solve design problems.

Engaging in Argument From Evidence

- Compare and refine arguments based on an evaluation of the evidence presented.

- Respectfully provide and receive critiques from peers about a proposed procedure, explanation, or model by citing relevant evidence and posing specific questions.

- Construct and/or support an argument with evidence, data, and/or a model.

- Use data to evaluate claims about cause and effect.

- Make a claim about the merit of a solution to a problem by citing relevant evidence about how it meets the criteria and constraints of the problem.

Obtaining, Evaluating, and Communicating Information

- Obtain and combine information from books and other reliable media to explain phenomena.

- Read and comprehend grade-appropriate complex texts and/or other reliable media to summarize and obtain scientific and technical ideas and describe how they are supported by evidence.

- Communicate scientific and/or technical information orally and/or in written formats, including various forms of media as well as tables, diagrams, and charts.

Continued

Table 4.1. (*continued*)

DISCIPLINARY CORE IDEAS

ESS2.A: Earth Materials and Systems

- Earth's major systems are the geosphere (solid and molten rock, soil, and sediments), the hydrosphere (water and ice), the atmosphere (air), and the biosphere (living things, including humans). These systems interact in multiple ways to affect Earth's surface materials and processes. The ocean supports a variety of ecosystems and organisms, shapes landforms, and influences climate. Winds and clouds in the atmosphere interact with the landforms to determine patterns of weather.

ESS2.C: The Roles of Water in Earth's Surface Processes

- Nearly all of Earth's available water is in the ocean. Most fresh water is in glaciers or underground; only a tiny fraction is in streams, lakes, wetlands, and the atmosphere.

ESS3.C: Human Impacts on Earth Systems

- Human activities in agriculture, industry, and everyday life have had major effects on land, vegetation, streams, oceans, air, and even outer space. But individuals and communities are doing things to help protect Earth's resources and environments.

LS2.A: Interdependent Relationships in Ecosystems

- Decomposition eventually restores (recycles) some materials back to the soil. Organisms can survive only in environments in which their particular needs are met. A healthy ecosystem is one in which multiple species of different types are each able to meet their needs in a relatively stable web of life. Newly introduced species can damage the balance of an ecosystem.

LS2.B: Cycles of Matter and Energy Transfer in Ecosystems

- Matter cycles between air and soil and among plants, animals, and microbes as these organisms live and die. Organisms obtain gases and water from the environment and release waste matter (gas, liquid, or solid) back into the environment.

CROSSCUTTING CONCEPTS

Energy and Matter

- Energy can be transferred in various ways and between objects.

- Matter is transported into, out of, and within systems.

Scale, Proportion, and Quantity

- Standard units are used to measure and describe physical quantities such as weight and volume.

Systems and System Models

- A system can be described in terms of its components and their interactions.

Continued

Table 4.1. (*continued*)

COMMON CORE STATE STANDARDS FOR MATHEMATICS

MATHEMATICAL PRACTICES

- MP1. Make sense of problems and persevere in solving them.
- MP2. Reason abstractly and quantitatively.
- MP3. Construct viable arguments and critique the reasoning of others.
- MP4. Model with mathematics.
- MP5. Use appropriate tools strategically.
- MP6. Attend to precision.

MATHEMATICAL CONTENT

- 5.MD.A.1. Convert among different-sized standard measurement units within a given measurement system, and use these conversions in solving multi-step, real world problems.
- 5.MD.C.3. Recognize volume as an attribute of solid figures and understand concepts of volume measurement.
- 5.MD.C.3.A. A cube with side length 1 unit, called a "unit cube," is said to have "one cubic unit" of volume and can be used to measure volume.
- 5.MD.C.4. Measure volumes by counting unit cubes, using cubic cm, cubic inches, cubic feet, and improvised units.
- 5.MD.C.5. Relate volume to the operations of multiplication and addition and solve real world and mathematical problems using volume.
- 5.MD.C.5.B. Apply the formulas $V = l \times w \times h$ and $V = b \times h$ for rectangular prisms to find volumes of right rectangular prisms with whole-number edge lengths in the context of solving real world and mathematical problems.
- 5.NBT.A.3.A. Read and write decimals to thousandths using base-ten numerals and number names.
- 5.NBT.A.4. Use place value understanding to round decimals to any place.
- 5.NBT.B.5. Fluently multiply multi-digit whole numbers using the standard algorithm.
- 5.NBT.B.7. Add, subtract, multiply, and divide decimals to hundredths, using concrete models or drawings and strategies based on place value, properties of operations, and/or the relationship between addition and subtraction; relate the strategy to a written method and explain the reasoning used.

Continued

Table 4.1. (*continued*)

COMMON CORE STATE STANDARDS FOR ENGLISH LANGUAGE ARTS

READING STANDARDS

- RI.5.1. Quote accurately from a text when explaining what the text says explicitly and when drawing inferences from the text.

- RI.5.4. Determine the meaning of general and domain-specific words and phrases in a text relevant to a *grade 5 topic or subject area*.

- RI.5.7. Draw on information from multiple print or digital sources, demonstrating the ability to locate an answer to a question quickly or to solve a problem efficiently.

- RI.5.8. Explain how an author uses reasons and evidence to support particular points in a text, identifying which reasons and evidence support which point(s).

- RI.5.9. Integrate information from several texts on the same topic in order to write or speak about the subject knowledgeably.

- RF.5.3. Know and apply grade-level phonics and word analysis skills in decoding words.

- RF.5.4.A. Read grade-level text with purpose and understanding.

- RF.5.4.B. Read grade-level prose and poetry orally with accuracy, appropriate rate, and expression on successive readings.

WRITING STANDARDS

- W.5.2. Write informative/explanatory texts to examine a topic and convey ideas and information clearly.

- W.5.2.B. Develop the topic with facts, definitions, concrete details, quotations, or other information and examples related to the topic.

- W.5.2.C. Link ideas within and across categories of information using words, phrases, and clauses (e.g., *in contrast, especially*).

- W.5.2.D. Use precise language and domain-specific vocabulary to inform about or explain the topic.

- W.5.4. Produce clear and coherent writing in which the development and organization are appropriate to task, purpose, and audience

- W.5.6. With some guidance and support from adults, use technology, including the internet, to produce and publish writing as well as to interact and collaborate with others; demonstrate sufficient command of keyboarding skills to type a minimum of two pages in a single sitting.

- W.5.7. Conduct short research projects that use several sources to build knowledge through investigation of different aspects of a topic.

- W.5.8. Recall relevant information from experiences or gather relevant information from print and digital sources; summarize or paraphrase information in notes and finished work, and provide a list of sources.

- W.5.9. Draw evidence from literary or informational texts to support analysis, reflection, and research.

Continued

Table 4.1. (*continued*)

SPEAKING AND LISTENING STANDARDS

- SL.5.1. Engage effectively in a range of collaborative discussions with diverse partners on grade 5 topics and texts, building on others' ideas and expressing their own clearly.

- SL.5.1.A. Come to discussions prepared, having read or studied required material; explicitly draw on that preparation and other information known about the topic to explore ideas under discussion.

- SL.5.1.B. Follow agreed-upon rules for discussions and carry out assigned roles.

- SL.5.1.C. Pose and respond to specific questions by making comments that contribute to the discussion and elaborate on the remarks of others.

- SL.5.1.D. Review key ideas expressed and draw conclusions in light of information and knowledge gained from the discussions.

- SL.5.4. Report on a topic or text or present an opinion, sequencing ideas logically and using appropriate facts and relevant, descriptive details to support main ideas or themes; speak clearly at an understandable pace.

- SL.5.5. Include multimedia components (e.g., graphics, sound) and visual displays in presentations when appropriate to enhance the development of main ideas or themes.

- SL.5.6. Adapt speech to a variety of contexts and tasks, using formal English when appropriate to task and situation.

FRAMEWORK FOR 21ST CENTURY LEARNING

- Interdisciplinary Themes: Health and Safety; Environmental Literacy; Science; Mathematics

- Learning and Innovation Skills: Creativity and Innovation; Critical Thinking and Problem Solving; Communication and Collaboration

- Information, Media, and Technology Skills: Information Literacy; Media Literacy

- Life and Career Skills: Flexibility and Adaptability; Initiative and Self-Direction; Social and Cross-Cultural Skills; Productivity and Accountability; Leadership and Responsibility

---

4

Table 4.2. Key Vocabulary for Lesson 1

Key Vocabulary	Definition
absorption	the process of soaking up a liquid
agriculture	the science or occupation concerned with preparing soil to grow crops and raising animals
architect	a person who creates structures and often supervises their construction
assumption	an idea accepted as true
civil engineer	a person who designs and maintains roads, bridges, dams, and similar structures
constant	a number used in calculations that does not change
cubic centimeter	a commonly used unit of volume that corresponds to the volume of a cube that measures 1 cm × 1 cm × 1 cm
desalination	the process used to remove salt from seawater to create freshwater that can be ingested by humans
formula	a calculation in written form; a group of mathematical symbols that express a relationship or solve a problem
fresh water	water that does not contain salt
groundwater	water found under Earth's surface in cracks and spaces between the soil, sand, and rocks
hydrologist	a person who gathers and evaluates data to predict droughts, create environmentally responsible regulations, and determine levels of pollutants
irrigation	a method of transporting water to crops to maximize the amount of yield from crops
layer	one thickness of something laid over another
meniscus	the curved (concave) upper surface of water in a container
milliliter	a measurement that is one-thousandth of a liter
molecule	the smallest unit of an object that still has the physical and chemical characteristics of the larger object
opaque	used to describe an object that one is not able to see through

Continued

Table 4.2. (*continued*)

Key Vocabulary	Definition
pi	a mathematical constant that represents the relationship between a circle's circumference and its diameter; represented by the symbol π
poetry	a kind of literature that uses words' meanings, sounds, and rhythms to create feelings and spark the imagination of readers
potable	safe to drink
precipitation	water in the form of rain, snow, sleet, or hail that falls from the sky
prototype	a model used to refine a final design
radius	the distance from the center of a circle to its outside edge
rain gauge	an instrument used to gather and measure the amount of rain (or other form of precipitation) that falls
rectangular solid	a three-dimensional object with six faces (four sides and two ends) that are rectangles
salt water	water that naturally contains salt, such as the ocean and the Great Salt Lake in Utah
surface water	water found on Earth's surface, such as that in lakes, rivers, and streams
volume	the amount of space that an object occupies, measured in cubic units
water management	a plan to capture and use water in ways that maximize its availability for human uses, such as providing drinking water and watering crops
watershed	an area or ridge of land that separates waters flowing to different rivers, basins, or seas

TEACHER BACKGROUND INFORMATION

In this module, students engage in ongoing consideration of rainwater management in homes and buildings. It is important to note that students' experiences will vary according to the setting in which they live. For example, students living in urban areas with high-rise apartment buildings may not be able to see gutters on buildings as easily as those who live in single-family homes, and students living in rural areas may not be familiar with storm sewer system components such as the storm drains commonly found in urban areas. In spite of these differences in context, all students should be able to observe some components of rainwater management around their homes.

Students will also consider irrigation for farm crops and gardens in this module. Students living in urban and suburban areas may lack firsthand experience with farms and irrigation systems; providing videos and images of farms, gardens, and irrigation systems may be particularly useful for these students. In addition, you may wish to connect the gardening concepts in the module to urban gardening initiatives in your area.

This lesson addresses volume measurement. This concept is essential for the main design challenge of the module, the Rainwater Roundup Challenge. Important components of this challenge include calculating the amount of water available, determining the amount of water needed for a garden, and designing a storage vessel to store rainwater.

Rainfall statistics are not provided in volume measurements, but rather, rain is measured in inches in the United States and in centimeters in Canada. Volume is, however, a consideration whenever water is captured, stored, drained, or transported. An hour's moderate rainfall may measure ¼ inch in a rain gauge. This seemingly small amount of precipitation translates to over 7,000 gallons of water on a surface the size of a football field.

Volume Calculations

Volume is measured in cubic units such as cubic centimeters, cubic meters, cubic inches, and cubic feet. Traditional measures of volume, such as milliliters, liters, or gallons, all have a cubic equivalent:

- 1 milliliter = 1 cubic centimeter

- 1 liter = 1,000 cubic centimeters

- 1 gallon = 231 cubic inches

The volume for a rectangular solid is calculated by multiplying length times width times height: $V = l \times w \times h$. The volume for a cylinder is calculated by multiplying the area of the base times height: $V = \pi \times r^2 \times h$.

Engineering Design Process

The engineering design process (EDP) is a series of steps that engineers follow when they are trying to solve a problem. The solution often involves designing a product (such as a machine or computer code) that needs to meet certain criteria or accomplish a particular task. The steps of the engineering process are as follows:

- *Define:* Define the problem. What is the problem? How have others approached it? Identify the requirements.

- *Learn:* Brainstorm possible solutions, and then research and find the best solution.

- *Plan:* Do research, list materials needed, and identify steps you will take.

- *Try:* Follow your plan and build a prototype.

- *Test:* Test the prototype. What works and what doesn't? What could you improve?

- *Decide:* Redesign to solve problems that came up in testing, and then retest.

Engineers do not always follow the EDP steps in order. It is common to design something, test it, find a problem, and then go back to an earlier step to make a modification or change the design entirely. This way of working is called *iteration.*

Rain Gauge Measurements

Rain gauges work on the principle of sampling. The opening for the rain gauge need not be large to measure the rain falling in the area where it is deployed; however, there are two important requirements:

- The instrument should not be obstructed by trees, overhangs, or anything else that might block the rainfall.

- The scale in the gauge is determined by the ratio of the size of the vessel to the size of the mouth.

Students may propose putting a large funnel on the top of the rain gauge, assuming this will allow them to collect a better representation. However, a collection opening (the funnel) that is much wider than the container (the straight tube) would allow a greater amount of water into the tube and distort the measurement. A rain gauge with the mouth the same size as the body has a 1 to 1 scale: 1 inch of water in the tube equals 1 inch of rainfall. A rain gauge with the mouth wider than the body would require that each inch marking on the scale be greater than an inch to render an accurate measurement. See Figure 4.1 for a visual representation.

With a rain gauge, the concern is not with the volume of the water in the gauge. Instead, a rain gauge measures the amount of precipitation that falls in a particular spot. This is the depth of the rain that would have accumulated on Earth's surface if it did not run off or soak into the ground.

Water

Water is an important molecule because it supports so many different forms of life. Water is an essential component of the cells of all living things. More than half (about 60%) of the adult human body is water. Without water, life on Earth would cease; therefore, it is important for students to understand how much is available, where it comes from, and what steps people can take to ensure that there is potable (drinkable) water to sustain life in the future.

When looking at a globe, students may notice that most of Earth is covered in water, but they may not be aware that 97% of the water is not potable, and only one-half of

1% is readily available in surface waters (rivers, lakes, and streams). This module provides an opportunity to discuss with students that potable water is a scarce resource. The article "Water, An Endangered Global Resource" on the Ahead of the Herd website at *http://aheadoftheherd. com/Newsletter/2012/Water-An-Endangered-Global-Resource.htm* provides an overview of global water usage and scarcity concerns. Innovations in technology infrastructures will be essential to overcome water shortages.

Figure 4.1. Rain Gauge Collection and Measurement With and Without Funnel

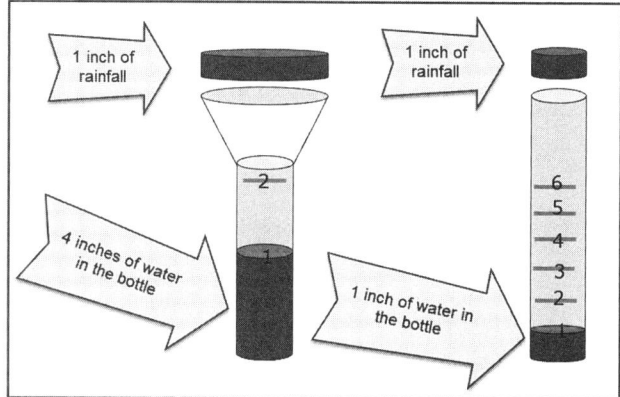

STEM Career Connections

You may wish to introduce students to the following STEM career connections during this module (adapted from Capobianco et al. 2015):

Hydrologist. Hydrologists gather and evaluate data to predict droughts, create environmentally responsible water regulations, and determine levels of pollutants using satellite instruments and computerized models showing how water moves above, on, and under Earth's surface. Hydrologists must have excellent analytical, computer, and interpersonal skills and must be able to endure strenuous outdoor activities in remote locations. It is likely that hydrology will be an increasingly important field in the 21st century because of concerns about water shortages. Hydrologists have bachelor's degrees and often master's degrees.

Civil Engineer. A civil engineer designs infrastructure used by the general public, such as roads, bridges, buildings, and systems for supplying water and treating sewage. Civil engineers' projects improve travel and commerce, provide people with safe drinking water and sanitation, and protect communities from earthquakes and floods. Civil engineers are creative, detail-oriented people who have good communication skills. Civil engineers have bachelor's degrees and often master's degrees.

Architect. Architects blend art and science by designing structures such as houses, schools, stores, offices, places of worship, museums, sports stadiums, theaters, and more. They must consider not only the structure's appearance, but also its safety, function, environmental impact, and cost. They often participate in every phase of design, from the initial vision to its completion. Architects are creative and have excellent spatial skills, as well as the ability to communicate and work effectively alone or in groups. Architects have bachelor's degrees and often master's degrees.

Landscape Architect. Landscape architects create landscape design plans for public areas such as parks, play spaces (e.g., skateboard parks), and the grounds of public buildings. The goal of a landscape architect is to create a design that is functional yet well balanced with nature, an environment where people feel happy and comfortable. Landscape architects must be able to read and comprehend technical reports, visualize what they read, and communicate information to their clients. Many landscape architects have college degrees, which are recommended but not required to enter the field, although a degree is needed to advance.

The Bureau of Labor Statistics' *Occupational Outlook Handbook* at *www.bls.gov/ooh/ home.htm* provides overviews of these and numerous other STEM careers.

Poetry

Students will investigate and create poetry about rain in this lesson. The Pen and the Pad website provides an overview of poetry for elementary age students at *https://penandthepad. com/five-types-poems-elementary-school-12137596.html.* A number of websites also provide searchable databases for poems, including the following:

- Academy of American Poets, *www.poets.org*

- Poem Hunter, *www.poemhunter.com/poems/rain*

- Poetry Soup, *www.poetrysoup.com/poems/water*

COMMON MISCONCEPTIONS

Students will have various types of prior knowledge about the concepts introduced in this lesson. Table 4.3 outlines some common misconceptions students may have concerning these concepts. Because of the breadth of students' experiences, it is not possible to anticipate every misconception that students may bring as they approach this lesson. Incorrect or inaccurate prior understanding of concepts can influence student learning in the future, however, so it is important to be alert to misconceptions such as those presented in the table.

Table 4.3. Common Misconceptions About the Concepts in Lesson 1

Topic	Student Misconception	Explanation
Earth science	Only water on Earth's surface is used for human purposes.	Groundwater is an important water source for humans and supplies about a third of the water used in people's homes and businesses.
	Groundwater and surface water are not connected.	The entire hydrosphere is connected because of Earth's hydrologic cycle. Water passes constantly through all three parts of the cycle.
	Water evaporates only from large bodies of water like oceans and lakes.	Water evaporates from any surface, including plant leaves, people's skin, and soil.
Measurement	Rainfall is measured using liquid volume units.	Rainfall is measured using the number of inches of rain that fall on a surface where the rain is not absorbed by Earth's surface and does not run off.
	Very thin objects, such as paper or tape, have no volume, since their thickness is difficult to measure.	All matter takes up space (i.e., has volume) and has mass (weight).
	Volume and capacity are the same thing; if an object has no capacity, it has no volume.	Volume is the measure of the amount of space something takes up; all matter has volume. Capacity is the ability of an object to hold volume; only things shaped so that other things can fit inside them (e.g., a cup) have capacity.
	The relationship between the shape and size of a rain gauge container and the size of its mouth does not matter.	The size of the mouth of a rain gauge influences how much rain is collected and is a critical feature of the gauge. Regardless of the size and shape of the body of the rain gauge container, the scale of the gauge must reflect the size of the mouth.

PREPARATION FOR LESSON 1

Review the Teacher Background Information (p. 60); assemble the materials for the lesson; make copies of the student handouts; and preview the videos, slideshow, spreadsheet, and other websites recommended in the Learning Components section below. Have your students set up their STEM Research Notebooks (see pp. 24–26 for discussion and student instruction handout).

Be prepared to group students into design teams of three or four. These teams will work together throughout this five-week module.

Visit your local library to compile a collection of fiction and nonfiction books about water and sources of water (rivers, lakes, streams, glaciers, oceans), as well as the water cycle. The librarian may put together a collection for you if you call ahead. A list of suggested books for the module is provided on page 82.

If you wish to have students conduct online research about watershed protection in social studies, reserve the computer lab in advance, and identify several websites that will provide students with information about the watershed in the community, human activities that might affect it, and measures being taken to protect the watershed. Alternatively, you may wish to provide printed resources to students on this topic. Your local wastewater management organization may be able to provide you with information.

Decide on three types of poetry you want students to focus on for the lesson (for example, acrostic poetry, narrative poetry, and haiku), and have examples of each type on hand for the class to read aloud together.

For the introductory activity, create a bar graph on the board with spaces for six bars on the horizontal axis and with a scale of 0% to 100% on the vertical axis. The graph should be large enough so that students will be able to distinguish quantities as small as 1%.

Measure the height and width of each wall of the room (ignore windows and door openings for the purposes of this activity), and have these measurements available for students to use in mathematics class.

LEARNING COMPONENTS
Introductory Activity/Engagement

Connection to the Challenge: Begin each day of this lesson by directing students' attention to the driving question for the module and challenge: How can we use what we know about rainfall to design a system to provide water to a garden? Hold a brief student discussion of student prior knowledge on this topic, creating a class list of key ideas on chart paper, or you may wish to have students create a notebook entry with this information.

Driving Question for Lesson 1: How does geographic location affect annual precipitation, and how is precipitation measured?

Mathematics Class and Science, ELA, and Social Studies Connections: Tell students that their challenge in this module is to devise a way to capture and reuse rainwater. Hold a class discussion about the role water plays in supporting life on Earth, and then hold an interactive group discussion to introduce the idea that freshwater resources are limited.

Use the Google slideshow "Water and Life" at *http://tinyurl.com/WaterAndLife* to introduce students to the importance of water via an interactive group discussion. Stop the slideshow at the "Where Does Water Come From?" slide. Focus on the following concepts during the discussion, pausing to allow students to answer the questions as they are posed:

- *Favorite Things.* Have students talk about the topics listed, prompting them to talk about favorite family members, foods, games, and the other things shown on the slides. Ask them if they can give examples of other favorite things that are not included on the slides.

- *Living or Nonliving?* Walk through the pictures and other ideas students gave during the Favorite Things discussion, asking about each, Is it living or nonliving? It is likely that most of the students' favorite things are living or that they originated from living things (e.g., food). The goal is to build the sense that virtually everything that enriches and supports life comes from other living things, whether animals (including humans) or plants.

- *What Is Needed for Life? Slide 1.* Introduce Maslow's hierarchy of needs. Have students discuss where each of the listed favorite items belongs on the chart.

- *What Is Needed for Life? Slide 2.* The placement of some items shown may differ from what students selected. It's important that students understand that food, air, water, and sleep are part of the foundational "physical" layer of the needs hierarchy. Then, show students how each layer builds on the one below it. Emphasize that no life can survive without food and water, and that all food (plant and animal) depends on water.

- *Water—Foundation of Life.* Have students discuss their thoughts about this statement by Pope Francis: "Water is the most essential element for life, and the future of humanity depends on our capacity to guard it and share it." Ask the following questions:

  - What is humanity?

  - How important is water to life?

- What should we guard water from?

- Who should we share water with?

- *Where Does Water Come From?* Cycle through the images (working from left to right), and question students to create curiosity about the origin of water by asking, "Where does the water originate?" Students should understand that most of the fresh water that supports life on Earth falls from the sky and returns to the sky in a cycle that has been repeating for millennia.

STEM Research Notebook Prompt

Have students add an entry in their STEM Research Notebooks listing sources of water.

Next, introduce the design challenge by reading the scenario presented on the Rainwater Roundup student handout (p. 84). Tell the class that to solve the module challenge, they will first need to learn about water and will begin by investigating where water is stored on Earth. List the following water sources on chart paper: oceans, groundwater, lakes, ice, swamps, and rivers. Hold a class discussion about each to ensure that students understand what these sources are. Have students create a bar graph in their STEAM Research Notebooks, labeling each source of water on the horizontal axis and percentages from 0 to 100 on the vertical axis. Ask students to predict percentages for each source and to represent these on their bar graphs, reminding students that the total of the bars they create cannot exceed 100%.

Using a permanent marker, label each of the five cups with the name of a source of water (excluding ocean water). Use the beaker and eyedropper to measure out the colored water into the plastic cups, leaving the remaining water in the pitcher to represent the ocean:

- Rivers: 0.01 ml (5 drops)

- Swamps: 0.02 ml (10 drops)

- Lakes: 0.08 ml (40 drops)

- Groundwater: 9.0 ml

- Ice: 20.6 ml

- Ocean: 970 ml (all the remaining water)

After you have distributed the water into the cups, label the pitcher as "ocean."

Next, add the water source labels to the horizontal axis of the bar graph you created on the board (see Preparation for Lesson 1, p. 66). Ask students to estimate how tall the bars for each water source should be based upon the water distribution they observe in the cups and pitcher, and create a bar chart on the board using the following percentages: ocean water, 97%; ice, 2.1%; groundwater, 0.9%; lakes, swamps, and rivers, each less than 0.009%.

Then, have the class watch as you pour a generous amount of salt into the ocean water. Introduce the term *potable water* and explain that while the oceans contain most of our planet's water (97%), it is not potable because it is much too salty to use as drinking water. Desalination processes can be used to remove salt, but they are very expensive and use a lot of energy. The limited amount of fresh water on Earth must support an ever-expanding population of humans, plants (agriculture), and animal life.

Review the additional data about Earth's water distribution found on the Google spreadsheet at *http://tinyurl.com/WaterDistribution*. Hold a class discussion about the data, pointing out the features of the graphs, including the titles and the visual representation of data. Ask students to share their ideas about the benefits of using visual representations like circle graphs to share data (e.g., makes it easier to understand, allows the reader to notice patterns). Hold a class discussion about the implications of the distribution of fresh water on Earth, asking questions such as the following:

- Where is most of Earth's fresh water stored?

- Is this fresh water easily usable for human needs?

- Are you surprised about the distribution of fresh water?

- Do you think we should protect our groundwater and lakes?

Activity/Exploration

Mathematics Class: Remind students that their challenge is to create a rainwater-recycling system. Ask students to share their ideas about how water is usually measured. Next, ask students for their ideas about how we measure the amount of space water takes up (volume) and the units we use for volume (cubic inches or cubic centimeters). Point out that the metric system uses cubic centimeters and the standard system in the United States uses cubic inches. Also point out that liquids are often measured in liters (1,000 cubic centimeters) or gallons (231 cubic inches) and that for small amounts of liquid, we use milliliters or ounces.

Show students an online weather forecast that shows inches of precipitation, such as the Seattle, Washington, forecast found on the Weather Underground website at *www.wunderground.com/weather/us/wa/seattle*. Ask students to identify how the amount of rain is expressed (in inches) and ask them to share their ideas about why amounts of rain are described in this manner. Point out to students that by collecting rain in a container and measuring it in inches, we can arrive at a standard measurement for rain. Introduce the concept of a rain gauge as a tool to measure the amount of rain. Make sure students

understand that if we tried to measure the total amount of rain that falls on a piece of land without using a rain gauge, we would have to account for the area of the piece of land, the absorption of the soil, and how the water runs off the land. Ask students if the rain that falls on Earth has volume (yes, but it can be difficult to measure because of the factors mentioned above). Tell them that using a rain gauge can tell us the volume of rain that falls in an area. As a class, use the United States Geological Survey rainfall calculator at *water.usgs.gov/edu/activity-howmuchrain.html* to explore the relationship between inches of rain and volume of rain.

Next, tell students that they are going to be challenged to calculate volume for an object that is difficult to measure in all three dimensions: a playing card. They will learn that even very thin objects have volume. As with all layers, thin or thick, rainfall has volume, and we calculate rainfall volume by measuring the depth of a layer of captured rain. Once the layer's depth and the area of the surface is known, volume of rainfall on a surface can be calculated, whether the surface is a playground or a city. Introduce the activity with the demonstration below.

Demonstration to Review Volume

Shine a flashlight into an opaque coffee mug. Ask students to help by shining some additional flashlights into the mug. Tell them you want to fill the mug up with light, so much that you can't add any more. Hold a discussion in which you guide students to explain why this could never happen (because light is a form of energy, and energy does not take up space).

Next, pour water from a pitcher into the coffee mug. Tell students that you intend to put all the water from the pitcher into the mug. As the mug fills, it will become obvious why this is impossible. Guide students to explain why you cannot keep adding water to the mug (water takes up space, and you will run out of space in the mug). When the mug is full, it cannot hold any more, and the water will overflow. Ask the class what characteristic of water made it impossible to pour all the water from the pitcher into the coffee mug. Guide students to understand that water is matter, and that matter takes up space and has mass. Remind students that volume can be measured in different units. Have students share their ideas of the units used to measure volume, creating a class list. Tell students that they are going to create their own reference guide for unit conversions before they work on the Playing Card Challenge.

Have students work in teams to create a conversion table for measuring vessels you provide, using the Volume Conversion Table handout (p. 87). They should derive these conversions by using measurement and mathematics, not by looking them up on the internet or using a conversion calculator. Following are the conversions for the units on the handout:

- liter to milliliters (1 L = 1,000 ml)

- liter to cubic centimeters (1 L = 1,000 cm³)

- cubic meter to liters (1 m³ = 1,000 L)

- quart to milliliters (1 qt. = 946 ml)

- cup to milliliters (1 cup = 236 ml)

- quart to cups (1 qt. = 4 cups)

- cubic inch to cubic centimeter (1 inch³ = 16.4 cm³)

- cubic centimeter to cubic millimeter (1 cm³ = 1,000 mm³)

The conversion for liters to cubic centimeters may present a problem for students who don't know that 1 ml = 1 cm³. You should decide whether to give this equivalency to students or let them wrestle with calculating it, most likely by measuring the 1,000 ml beaker and using the formula for the volume of a cylinder: $V = \pi \times r^2 \times h$.

Playing Card Challenge

Pass out copies of the Playing Card Challenge student handout (p. 89). Introduce the activity by holding up a paperback book and asking students to describe the book's physical properties (shape, size, measurements). Elicit the term *rectangular solid* as students offer descriptions, and take a minute to review the formula for finding the volume of a rectangular solid: volume in cubic units = length × width × height ($V = l \times w \times h$).

Have students sketch a picture of a book in their STEM Research Notebooks, then have them work in pairs to discuss how to find the volume of a book. Challenge students to recall the formula for the volume of a rectangular solid. Ask the students to provide some examples of other rectangular solids in the world outside the mathematics classroom. Ask them to give an example of a rectangular solid that may be too small to have volume. Discuss their proposals.

Hold up a playing card and ask how much space the card takes up. Explain that this is more challenging than determining the volume of the book, because a typical classroom does not have a measuring tool suitable to determine the thickness of the card.

Show a 1 cm math cube or 1 cm game cube (die). Walk around the class showing the cube. Ask students to consider the volume of the playing card and the cube. Create a tally chart on the whiteboard with two columns: greater than (>) 1 cubic centimeter and less than or equal to (≤) 1 cubic centimeter. Take a poll, asking students whether they think the volume of the playing card is greater than 1 cubic centimeter or less than or equal to 1 cubic centimeter. Allow students to provide reasons to support their thinking.

Next, have students work in their design teams to brainstorm ideas for finding the volume of a single playing card. If the playing card has rounded corners, tell students to ignore this for their calculations.

One possible solution is to measure the thickness of the whole deck, calculate the volume of the deck as a rectangular solid, and then divide this by the number of cards in the deck. The volume of the card (ignoring rounded corners) should calculate to around 1.5 cubic centimeters, although the result may vary depending on the size of the cards. Here are the calculations for a deck of cards that is 2 ½ × 3 ½ inches:

- $V = l \times w \times h = 89$ mm $\times 64$ mm $\times 14$ mm $= 79{,}744$ cubic millimeters

- 79,744 divided by 1,000 cubic mm per cubic cm = 79.7 cubic centimeters

- 79.7 divided by 52 cards in the deck = 1.53 cubic centimeters per card

Alternatively, students may determine the number of cards in a 1 cm stack, which will change the calculation to the following:

- $V = l \times w \times h = 89$ mm $\times 64$ mm $\times 10$ mm $= 56{,}960$ cubic millimeters

- 56,960 divided by 1,000 cubic mm per cubic cm = 57.0 cubic centimeters

- 57.0 divided by 37 cards (stacked to get 1 cm) = 1.54 cubic centimeters per card

STEM Research Notebook Prompt

Describe the process your team used to find the volume of the card. Sketch the card as a rectangular solid. Label the dimensions as shown for the cube on the handout.

Science Connection: Students are challenged to use the EDP to design a rain gauge. Introduce the EDP to students and review the steps (see the Teacher Background Information section on p. 61).

Rain Gauge Design Challenge

Tell students that they will work in teams of three or four to design and build a rain gauge. They will work in these same design teams to complete the module challenge. In addition to designing and building rain gauges, teams will create spreadsheets or tables to record rainfall data. Each team will deploy its rain gauge at a unique location on the school campus to measure rainfall over the duration of the module.

Remind students that although water is usually measured using volume, rainfall is measured in units of height. Ask students these questions:

- What device is used to measure rainfall?

- What does a rain gauge measure? (The depth of rain that would accumulate in an area the size of its mouth if water could not run off or soak in)

- Do the shape and size of the container affect the height of the water collected in a rain gauge? (Yes, the size of the container affects the volume of water collected; a rain gauge measures the height of the water collected so the scale used to measure the height of that water will vary accordingly if the mouth size of the container used to collect the rainwater is different from the collection area.)

Next, show students two liquid measuring containers with two different diameters, such as a 1,000 ml beaker and a 1,000 ml graduated cylinder. Ask students to imagine that they had a big rainstorm in which 2 inches of water fell and answer the following questions:

- Would the height of the water in each of these containers be the same? (Yes)

- Would the readings on the beaker and cylinder be the same? (No)

- What would we have to do to these containers to make them into rain gauges? (Change the measurement scale to inches)

Review the garden scenario and the design challenge for the module, reminding students that they have been challenged to devise a way to collect rainwater to use to water a garden. Display the small pitcher, and ask students whether this could solve the problem of capturing rainwater. The students should respond that the pitcher is too small to collect enough water for a garden. Display the big watering can as an alternative. The students should note that although the watering can is larger, it would not collect enough water to water a garden. Ask the class to respond to the following questions:

- What facts can you cite to prove that the pitcher and watering can are not good solutions to capture and reuse rainwater?

- What information would you need to design the kind of water collection system needed in the Rainwater Roundup scenario?

- How could you get that information?

In the course of the discussion, students should mention the need for information about amounts of rainfall. If they do not, ask guiding questions. Introduce the Rain Gauge Design Challenge by handing out and reviewing the Rain Gauge EDP student handout. Next, hand out and review the Rain Gauge Design Challenge Rubric. Tell students that each rain gauge needs to meet certain requirements. Rain gauges must

- be weatherproof,

- measure rainfall in centimeters,

- have a permanent scale mounted on the inside or outside,

- have a scale that is calibrated to measure centimeters of rain on an area equal to the size of the mouth of the gauge (see the Rain Gauge Measurements section on p. 62), and

- be mounted or secured to prevent breakage.

In addition, students should provide weatherproof signage (for example, a laminated sign or a sign encased in a resealable plastic bag sealed with duct tape) to indicate its purpose to other students in the school.

After teams have constructed their rain gauges, each team should present its rain gauge to one other team and ask for feedback about the design and function of the rain gauge. Then, students on each team should work together to determine the best location for the rain gauge their team created, using their experiences with their own schoolyard. To avoid having all rain gauges placed in the same area, assign each team an area where they may place their rain gauges (e.g., at the front of the school building; on the north, south, east, or west side of the school building; on the playground; beside the parking lot).

Next, have each team create a spreadsheet or table for data collected from the rain gauge, including the dates of the readings, location of the rain gauge, and the calculation of the previous 24 hours of rainfall for each date. Days with no rainfall should also be included. Teams should create a proposal for how and when the data should be gathered. In addition, they should come up with a strategy for no-school days such as weekends and holidays.

STEM Research Notebook Prompt

Have students respond to one or more of the following reflection questions in their notebooks:

- *Describe how the rain gauge design changed as you worked through the EDP.*

- *What additional or different materials would your team have used if they were available?*

- *If you were going to make a rain gauge to sell, how would you change the design?*

- *How did your team's rain gauge differ from those built by other teams?*

- *How did your team's rain gauge change after receiving feedback from another team?*

ELA Connection: Students may use this time to design a cover page for their STEM Research Notebooks using the title Rainwater Roundup Challenge. Allow students to use art materials and add pictures they have printed out, word art, or even hand-drawn cartoon images. Students should add the title to the table of contents in the front of the notebook.

Next, use the last four slides of the "Water and Life" slideshow at *http://tinyurl.com/ WaterAndLife* to read aloud the poem *After the Rain*, by Valerie Dohren. Launch a class discussion about rain's benefits to living things, and the feelings students experience on rainy days compared with sunny days.

Ask students to share their ideas about what poetry is and create a class definition. Next, ask students how poetry compares with the kind of texts they usually read in science class. Create a class chart of student ideas about how poetry and science writing are the same and different. Ask students if they could use a poem to talk about a science idea. What kind of poem could they use?

Introduce the idea that nature is often the subject of poetry. Ask students to share their ideas about why this might be. As a class, look through the titles in a nature poetry book (such as *National Geographic Book of Nature Poetry: More Than 200 Poems With Photographs That Float, Zoom, and Bloom!* by J. Patrick Lewis) or on a poetry website that focuses on nature (such as the Academy of American Poets' "Nature Poems" page at *www.poets.org/ poetsorg/poems?field_poem_themes_tid=1226*). Ask students to share their ideas about what the titles of poems tell the reader about the topic of the poems, and create a class list of the topics of the nature poems (e.g., animals, weather, bodies of water, flowers). Have students create a section in their STEM Research Notebooks to record poetry they create and their reactions to poetry they read throughout the module.

Throughout the module, encourage students to read fiction and nonfiction material dealing with water, sources of water (surface and underground water, clouds), how water is used, and the importance of water in agriculture. (See p. 82 for a list of suggested books.)

Social Studies Connection: In social studies, students learn about groundwater, watersheds, and water management.

STEM Research Notebook Prompt

Ask students to think about what groundwater looks like, where it is found, and how it gets there, and then create labeled drawings of their ideas in their STEM Research Notebooks.

Next, hold a class discussion on the following:

- Are any of the following statements true? Most groundwater is found … in a huge lake under the ground (No); in an underground river far below the soil (No); in between particles of sand or gravel (Yes); in the solid bedrock underground (No); on Earth's surface in rivers and lakes (No).

- Does groundwater move underground? (Yes)

- Is surface water connected to groundwater? (Yes)

Watershed Model

Tell students that water scarcity has a huge impact on food production. Without water, people do not have a means of irrigating crops to provide food for the population. Agriculture (70% of global water usage) is constantly competing with domestic, industrial, and environmental uses. To minimize this problem, people have developed methods of water management.

Introduce the idea that one method of water management is irrigating fields. Irrigation is a way to transport water to maximize the yield of crops produced in a region. Irrigation may involve transporting water from nearby surface water sources such as rivers or lakes to fields, or it may involve capturing rainwater. Point out to students that not all irrigation systems use water in an effective manner and may add too much water to crops, so that the excess water runs off and into nearby streams and other waterways. Ask students why this might be a problem. (When the water runs off, it carries soil and fertilizers back to the waterways; this can pollute and change the environment of surface water.)

Tell students that in the module challenge, they are going to create a plan for water management that involves capturing rain, which they can then use to water a garden. After investigating the need for water-capturing techniques for agriculture, students will investigate several rain-fed irrigation techniques in this social studies lesson. The methods include water-harvesting systems, rain catchment systems, and weirs or sand dams.

Introduce the term *watershed* and the idea of a watershed as a precipitation collector. Ask students to share what they know about watersheds with their team members. Use the USGS Water Science School's "What Is a Watershed?" web page at *https://water.usgs. gov/edu/watershed.html* to introduce the concepts associated with watersheds.

Tell students that they are going to create models of watersheds. Pass out a sheet of 8½ × 11 inch paper to each student. Instruct students to crumple the paper into a tight ball, then gently open the paper but don't flatten it out. Tell students this piece of paper represents a watershed. Ensure that each student has access to water-soluble brown, blue, red, and green markers, and instruct students as follows:

1. Have students use the *brown* marker to place a mark on all the highest points on their watershed (crumpled paper). Explain that the high points represent areas with a high elevation, such as mountains and hills.

2. Most bodies of water are located in lower elevations. Use the *blue* marker to mark the places on the paper where bodies of water might be located. Guide students to share what they remember about the creeks, rivers, and lakes they have visited. What signs are present in these areas that indicate a lower elevation?

3. Ask students to use the *red* marker to color two or three areas on the paper to represent human settlements. Discuss the impact these areas might have on the watershed (use of water for drinking, sanitation, lawn irrigation, potential sources of pollution).

4. Have students use the *green* marker for two or three agricultural areas where plants or animals could be raised. Ask students to discuss with their teams the needs of plants and animals and how these needs might have an impact on the watershed.

5. Finally, pass out misting spray bottles and tell students to lightly spray the finished paper terrains. The spray will represent rain falling into the watershed.

As a class, discuss how the water is traveling through the watershed system and students' observations, asking the following questions:

- What changes occurred to the model?

- Where did most of the water go?

- Did the water flow in a path? Where did it go?

- What happened to the human settlements? Were they in the path of a raging river or crumbling mountain?

- What happened to the agricultural areas? Would water flowing through these areas affect the other regions?

- What if the marker were a dangerous chemical used on crops? What would happen to living creatures along the path of the watershed?

- What actions could farmers take to protect water quality?

- How does this model demonstrate the interconnectedness of a watershed?

Explanation

Mathematics Class and Science Connection: Review volume measurement calculations for rectangular solids ($V = l \times w \times h$). One important concept that should be stressed to students is that all liquid measurements are volume measurements. In everyday life, students who assist in the household kitchen will be familiar with quarts, cups, teaspoons, and so forth. The volume conversion activity helped them equate the volume in a nonrectangular container with cubic volume units. Have several teams of students present their volume conversion charts to the class. As a class, discuss what volume units were used, and have students brainstorm ideas about when they might need to convert between these units.

Using a beaker or graduated cylinder filled with water, ask students to look at the water and decide whether it is flat on the surface. Students should note that the center is lower than the edges. Introduce the term *meniscus*. Tell students that the meniscus forms because water particles are attracted to the glass walls of the container. Ask students to share their ideas about how the meniscus influences our ability to measure water. Demonstrate to students the procedure of looking at the container at eye level and measuring the lowest surface of the meniscus to arrive at the most accurate measurement.

Ask students for their reactions to a weather report that tells them to expect ¼ inch of rain. Does this seem like a lot of rain or not very much? Point out to students that spread over a large surface or geographic area, a ¼-inch-thick layer of water is a very large volume. Tell them that in March 1982, 1 inch of rain fell on snowy ground and caused a historic flood in Fort Wayne, Indiana, that surrounded 2,000 homes and caused the evacuation of 9,000 people. Ask students to share their ideas about why 1 inch of rain was such a catastrophic event. If students are unable to generate ideas, emphasize the point that the ground was already covered with snow that was melting. Also point out to students that rain does not lie on the surface of the ground very long. The rain either soaks into the soil or runs off to drains or waterways. Ask students what they think will happen when the ground is so soaked it cannot hold more water (the rain runs off into drains and waterways) and what might happen to the waterways when the ground is soaked (the level of the water rises and may flood surrounding areas).

Next, have student teams present their solutions to the Playing Card Challenge to the class. Compare teams' strategies for finding the thickness of a playing card, and create a chart that lists all teams' answers.

ELA Connection: Introduce the idea to students that there are different types of poetry. Have students share their ideas about types of poetry they have encountered, creating a class list. Focus students' attention on two or three types of poems (for example, acrostic poems, narrative poems, and haiku), and read examples of each aloud to the class. Ask students to share their ideas about the similarities and differences among the types of poetry, recording student responses on a class chart. Emphasize to students that poetry does not have to rhyme. Formulate a class definition of poetry.

Social Studies Connection: Hold a class discussion about the watershed, asking students to share their ideas about why it is important to the overall supply of human water, what kinds of human activities might affect the watershed, and how it can be protected. Record student ideas on chart paper. Students should understand that watersheds supply water for drinking, business needs, and irrigation of crops, and they are also habitats for many plants and animals. Students should also understand that human activities that result in any sort of pollution can have an impact on the watershed, since water we use daily may eventually reach rivers, lakes, and streams. In addition, when people change the geological features of the land by, for example, mining or using fill dirt to create level areas for constructing buildings, this can change the way that water runs off the land.

Emphasize to students that farming has an important impact on the watershed. Since people involved in agriculture realize that water is necessary to grow crops, they may take measures to conserve and protect the water in their watershed. Following are some of the farming practices that are used to conserve water and protect the watershed:

- Planting grass strips between the crops and surface waters to slow the water as it runs off the field and trap potential pollutants before they reach the water source.

- Leaving plant remnants on the field after harvesting to reduce the amount of soil washed into the water source.

- Keeping manure (animal waste) contained to prevent its leakage into the groundwater and surface water sources.

Watershed Conservation Research and Place Mat Design

Have students conduct research, using the websites or printed resources you put together in advance (see Preparation for Lesson 1 on p. 66), to learn more about watersheds in their community.

- What are the watersheds in your community? (Where does the rain go?)

- What affects the watershed in your area?

- What can farmers and gardeners do to protect the watershed?

- What can families in the community do to protect the watershed?

Hand out the Watershed Place Mat Rubric and review it with students. Have each student create a place mat on an 11 × 17 inch sheet of paper to inform others about what they learned about watersheds, providing a section for their findings about each question above.

Elaboration/Application of Knowledge

Mathematics Class: Ask students to share their ideas for activities in their homes where volume measurements may be important (e.g., cooking). The playing card volume activity illustrated that even a thin layer has volume. Now, work together as a class to solve a volume problem related to painting. Provide a fictional scenario to students, telling them that they have been asked to paint their classroom. Ask students what they need to know to do this (how much paint to buy). Tell students that paint dries to a very thin layer that you will estimate at 0.1 mm. Provide students with the measurements of each wall (ignoring door and window openings). Have students work in teams, using their

volume conversion sheets and the volume formula for rectangular solids, to determine how much paint they would need to buy in order to paint the room.

Science Connection: Tell students that they will need to collect data from their rain gauges each time it rains. Ask students for their ideas about how they could collect and track those data (e.g., assign someone from the team to check the rain gauge daily, record the amount of rain in a chart). Ask students for their ideas about what data they should record about their rain gauges (e.g., date, time, amount of rain in gauge). Have student teams work together to create ways to record the data. Students may choose to use handwritten tables, spreadsheet software, or a form that they fill out periodically. After teams have created their data collection plans, have each team share its ideas with the class, discussing the features of each data collection plan, what students like about each, and improvements that could be made. Based on these discussions, formulate one plan for all teams to use to collect data from their rainwater gauges. For the remainder of the module, students should check the rain gauges regularly to take rain measurements and to be sure that the gauges are intact and operating properly.

Next, emphasize to students that engineers design systems to solve problems. Remind students that the module scenario takes place in a small California town under water use restrictions. The climate where their school is located may differ substantially from that of the fictional town of La Vieja. As a class, discuss how the climate where you live is the same as or different from that of La Vieja.

Ask students for their ideas about how engineers design things for environments that are different from the environment in which they live and work, emphasizing the need for research. Use as an example the engineers who built the Mars rovers. Ask students to share their ideas about what challenges engineers might have faced in building a vehicle for a planet they have never visited. Read aloud a book about Mars rovers, such as *Curiosity: The Story of a Mars Rover* by Markus Motum. After reading, have students share what they learned about how the engineers designed the rover. Ask students how they could use what they learned to design the rainwater-recycling system for the La Vieja garden.

ELA Connection: Have each student choose one of the types of poetry you discussed in this lesson. Hand out the Poetry Writing Rubric and review with students, telling them that they will use this handout to evaluate their own poems and their classmates' poems. Have each student create a poem relating to rain or watersheds. After students have completed their poems, have them read two classmates' poems and provide feedback using the Poetry Writing Rubric.

Social Studies Connection: Have several students share their place mats with the class. Review the concept of a watershed as a geographic area where water drains into a common body of water. The watershed includes all the plants, animals, and people that live in the area, as well as the nonliving components in the soil. Emphasize that since we all are part of a watershed, everything we do can potentially affect the surface water and groundwater that run through this system.

In the course of learning about water sources, students will learn that water sources are interconnected. Encourage students to find and take pictures of storm runoff, sewer grates, and signs posted along river and stream banks. Have the class brainstorm ideas for ways they can take an active part in protecting water sources. Introduce the idea of public service advertisements (PSAs), which are meant to raise community awareness about an issue. Tell students that often these efforts use a slogan, a short phrase that is memorable, to make their point. Provide some examples of slogans students may be familiar with (e.g., "Just do it"; "Snap, Crackle, Pop"; "Take a bite out of crime"; "Only YOU can prevent wildfires"). Ask students to share their ideas about what makes these slogans effective (e.g., they are short, they are memorable, they relate to the purpose of the product or message). Have each team work together to create a slogan that they think would be useful to provide a message to their community about the importance of watershed protection. Then, have each team present its slogan to the class, and have the class vote for one of the slogans to be the class slogan for watershed protection.

Evaluation/Assessment

Students may be assessed on the following performance tasks and other measures listed.

Performance Tasks

- Playing Card Challenge

- Rain Gauge Design Challenge Rubric

- Watershed Place Mat Rubric

- Volume Conversion Table

- Watershed Model activity

- Poetry Writing Rubric

Other Measures

- STEM Research Notebook entries and rubric

- Engagement in class activities and discussions

- Involvement in group work and discussions

INTERNET RESOURCES

"Water and Life" slideshow
- *http://tinyurl.com/WaterAndLife*

Water Distribution spreadsheet
- *http://tinyurl.com/WaterDistribution*

"Water, an Endangered Global Resource"
- *http://aheadoftheherd.com/Newsletter/2012/Water-An-Endangered-Global-Resource.htm*

Bureau of Labor Statistics' *Occupational Outlook Handbook*
- *www.bls.gov/ooh/home.htm*

Overview of poetry for elementary age students
- *https://penandthepad.com/five-types-poems-elementary-school-12137596.html*

Poetry databases
- *www.poets.org*

- *www.poemhunter.com/poems/rain*

- *www.poetrysoup.com/poems/water*

Weather Underground Seattle weather forecast
- *www.wunderground.com/weather/us/wa/seattle*

United States Geological Survey rainfall calculator
- *water.usgs.gov/edu/activity-howmuchrain.html*

Academy of American Poets' "Nature Poems"
- *www.poets.org/poetsorg/poems?field_poem_themes_tid=1226*

SUGGESTED BOOKS FOR THIS MODULE

- *A Drop Around the World,* by Barbara Shaw McKinney (Dawn Publications, 1998)

- *A Drop of Water: A Book of Science and Wonder,* by Walter Wick (Scholastic Press, 1997)

- *Clean Water for Elirose,* by Ariah Fine (Createspace Independent, 2010)

- *Did a Dinosaur Drink This Water?* by Robert E. Wells (Albert Whitman & Company, 2006)

- *Farmer George Plants a Nation,* by Peggy Thomas (Calkins Creek, 2013)

- *McElligot's Pool,* by Dr. Seuss (Random House Books for Young Readers, 1947)

- *One Well: The Story of Water on Earth,* by Rochelle Strauss and Rosemary Woods (Kids Can Press, 2007)

- *The Life and Times of a Drop of Water: The Water Cycle,* by Angela Royston (Raintree, 2005)

- *The Magic School Bus Wet All Over: A Book About the Water Cycle,* by Pat Relf (Scholastic Press, 1996)

- *The Water Cycle,* by Bobbie Kalman (Crabtree, 2006)

- *The Water Cycle: Evaporation, Condensation & Erosion,* by Rebecca Harman (Heinemann, 2005)

- *The Water Hole,* by Graeme Base (Harry N. Abrams, 2001)

- *Thomas Jefferson Grows a Nation,* by Peggy Thomas (Calkins Creek, 2015)

- *Water Can Be …,* by Laura Purdie Salas (Millbrook, 2014)

- *Water Dance,* by Thomas Locker (HMH Books for Young Readers, 2002)

- *Water Sources,* by Rebecca Olien (Capstone Press, 2016)

IMAGE CREDITS FOR LESSON 1

Figures and photographs (Owner: Pandaia Projects LLC. Used with permission.)
- Figure 4.1 (p. 63)

- Playing card photograph (p. 89)

REFERENCE

Capobianco, B. M., C. Parker, A. Laurier, and J. Rankin. 2015. The STEM Road Map for grades 3–5. In *STEM Road Map: A framework for integrated STEM education,* ed. C. C. Johnson, E. E. Peters-Burton, and T. J. Moore, 68–95. New York: Routledge. *www.routledge.com/ products/9781138804234.*

STUDENT HANDOUT, PAGE 1

RAINWATER ROUNDUP

Wilbur Wright Elementary is a neighborhood school in La Vieja, a sunny town in Southern California. The school is near a retirement home called Sunny Acres. Students walking to school often see residents of Sunny Acres, and two students have relatives who live there.

Jorge's great-grandmother, Anna, lives in Sunny Acres. Sometimes while walking to school, Jorge and his friends see Mamito Anna driving her golf cart along the front sidewalk near Bristol Boulevard. She jokes that La Vieja, which means "the old lady" in Spanish, was named after her. Her golf cart is usually crowded with art supplies: an easel, half-finished canvases, paints, and brushes. Mamito Anna says she likes to get up early and do her painting near Bristol Street because the morning light brings the whole world to life. Her favorite subjects to paint are people coming and going from the homes, shops, and buildings around La Vieja. One of Mamito Anna's paintings, a watercolor of children standing in front of the school, hangs in the principal's office. Anna likes to say that Jorge's father is one of the children in the picture and that the day the children posed for the picture was the only day Jorge's papa ever stood still.

Angie's grandfather lives in Sunny Acres, too. She never sees him on the walk to school because he spends his mornings in the flower garden. Grandpa Henry jokes that because the garden is near the dining hall, he spends his time in the Garden of Eatin'. He loves the garden, and his friends at Sunny Acres love the fresh flowers he uses to decorate the tables and desks throughout the retirement home.

One morning, Angie was surprised to see Grandpa Henry painting with Mamito Anna under a tree along Bristol Street. "Grandpa," she said, "I didn't know you liked to paint."

"Don't know if I'd call it painting," he grumbled. "It looks more like someone held up a canvas in the middle of a food fight."

"Shhh, Henry, don't be so hard on yourself," cooed Mamito Anna. "It's not bad for a beginner. We'll have you painting beautiful flower pictures before you know it."

NATIONAL SCIENCE TEACHERS ASSOCIATION

STUDENT HANDOUT, PAGE 2

RAINWATER ROUNDUP

"Flower pictures, Grandpa?" Angie looked at Grandpa's picture, trying to visualize it as flowers. "Why would you want to paint flower pictures when you have all those beautiful flowers in the garden?" she asked.

Grandpa Henry stopped painting and looked toward the dining hall. "There's not going to be any flowers in the garden anymore. We're turning it into a rock garden."

"A rock garden? Why would you take away the flowers? You love the flowers! Everyone loves the flowers!" protested Angie.

"It's the water shortage, dear," said Mamito Anna. "La Vieja and all the towns in the valley need to conserve water to make sure there is enough for people to use for drinking, eating, and washing. Watering with a hose is no longer allowed for lawns or gardens."

"Oh, Grandpa," Angie lamented. "This is the worst news ever!"

Grandpa Henry held up his painting. "What do you think, Anna? Should we hang this one up in the dining hall by the table where we always had the vase of daisies? This yellow and green splotch kind of looks like a daisy … if you turn the canvas sideways … and squint just right."

"Hmmm …," said Mamito Anna. "I'm not quite sure it's ready." She brightened and continued, "But you did say it looked like a food fight." She winked at Angie. "Perhaps it would fit right in."

Angie, Grandpa Henry, and Mamito Anna all laughed. Just then, Angie remembered a video she had seen in science class about water and conservation. It showed students building a rainwater-recycling system for a school garden in a desert town in Arizona. "I saw something last week about recycling rainwater for gardens," she said.

"Rainwater!" Grandpa Henry snorted. "It only rains three months out of the year around here. What are you going to do when it doesn't rain for weeks?"

STUDENT HANDOUT, PAGE 3

RAINWATER ROUNDUP

"Grandpa, I'm serious! People are growing flowers in the desert with recycled rainwater," Angie exclaimed. "There's no time to lose! I have to get to school!" Off she went. As she ran, Angie called back over her shoulder, "Don't let them dig up the garden yet, Grandpa! I have an idea."

"She's a smart one, that girl," said Grandpa Henry as he and Mamito Anna watched Angie race away, her purple backpack bouncing around with every step. "Wouldn't it be something if she could help us!"

THE RAINWATER ROUNDUP CHALLENGE

Sunny Acres has two main buildings, each about the same size as Wilbur Wright Elementary School. The garden is 8 m × 10 m (or about 26 feet × 32 feet) and is located between the buildings in an open courtyard. The parking lot is similar in size to your school playground. Your challenge is to help Wilbur Wright save the flower garden at Sunny Acres. Using your school building and grounds as a design lab, devise a method to capture and reuse rainwater around the school. With your research, formulate a proposal that you can bring to the residents and leaders at Sunny Acres about how they might be able to save the flower garden.

Team Name: _____   Date: _____

STUDENT HANDOUT, PAGE 1

VOLUME CONVERSION TABLE

In this investigation, you will create a conversion table for common volume measurements. You have received several measuring containers. Some are used to measure liquids in the metric system. Others are used to measure liquids in the American standard system.

INSTRUCTIONS

Complete the table to show how many of each conversion unit are required to be equal to the original unit. Each conversion should be rounded to the nearest unit except when converting 1 cubic inch to cubic centimeters, where you should round to the nearest tenth. The first two units are provided. In the Method column, indicate *M* for measure, *C* for calculate, or *MC* for measure and calculate.

Original Unit	How Many?	Conversion Unit	Method
1 inch	2.5	centimeters	given
1 gallon	4	quarts	given
1 liter		milliliters	
1 liter		cubic centimeters	
1 cubic meter		liters	
1 quart		milliliters	
1 cup		milliliters	
1 quart		cups	
1 cubic inch		cubic centimeters	
1 cubic centimeter		cubic millimeters	

Team Name: _____   Date: _____

VOLUME CONVERSION TABLE

STEM RESEARCH NOTEBOOK

In your STEM Research Notebook, explain the process you used to find the conversion factors.

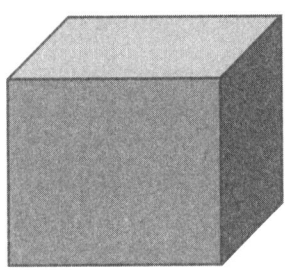

Create three-dimensional drawings of two containers that could hold exactly 1 liter each. You may choose any shapes you wish, such as a cube, a cylinder, or a rectangular container.

- Label your containers. Each container should have different dimensions.

- Use your conversion factors to find the dimensions of the containers.

Team Name: _____   Date: _____

PLAYING CARD CHALLENGE

Calculate the volume of a single playing card in cubic centimeters.

INSTRUCTIONS

Your team has received a deck of cards and a metric ruler. Calculate the volume of a single card to the nearest tenth of a cubic centimeter. Disregard the rounded corners.

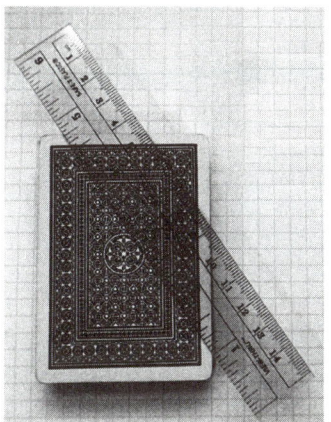

Describe the Process Used:

Data:

Results:

1 cm

1 cm

1 cm

Was the volume greater than, less than, or equal to the volume of the math cube or game cube?

Volume of One Playing Card: _____

RAIN GAUGE EDP

Identify the Problem

Build a rain gauge to measure rainfall in centimeters. It should be weatherproof and able to last outside for a month. You must also include a notice about the purpose of the equipment and whom to contact in case of problems.

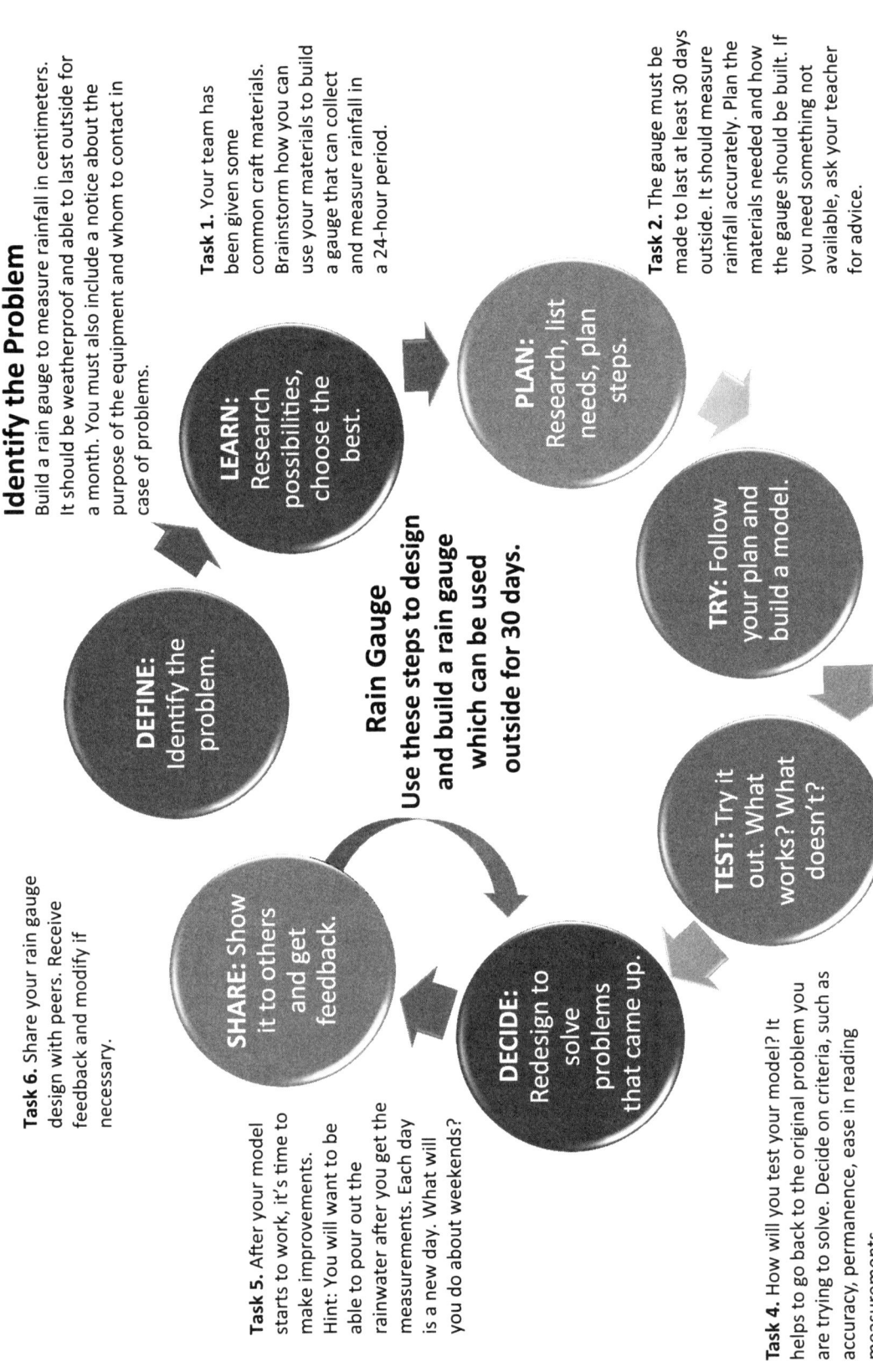

Rain Gauge
Use these steps to design and build a rain gauge which can be used outside for 30 days.

DEFINE: Identify the problem.

LEARN: Research possibilities, choose the best.

PLAN: Research, list needs, plan steps.

TRY: Follow your plan and build a model.

TEST: Try it out. What works? What doesn't?

DECIDE: Redesign to solve problems that came up.

SHARE: Show it to others and get feedback.

Task 1. Your team has been given some common craft materials. Brainstorm how you can use your materials to build a gauge that can collect and measure rainfall in a 24-hour period.

Task 2. The gauge must be made to last at least 30 days outside. It should measure rainfall accurately. Plan the materials needed and how the gauge should be built. If you need something not available, ask your teacher for advice.

Task 3. Get to work building your model using the plan you made.

Task 4. How will you test your model? It helps to go back to the original problem you are trying to solve. Decide on criteria, such as accuracy, permanence, ease in reading measurements.

Task 5. After your model starts to work, it's time to make improvements. Hint: You will want to be able to pour out the rainwater after you get the measurements. Each day is a new day. What will you do about weekends?

Task 6. Share your rain gauge design with peers. Receive feedback and modify if necessary.

	STEM Research Notebook Rubric
	Name: _____
10	**OUTSTANDING** • The writing goes beyond the basic requirements and shows in-depth understanding of concepts. • The work shows in-depth reflection throughout the learning process. • The notebook has all the components expected, including dates and labels on each page. • All pages are numbered properly, with odd numbers on the right and even numbers on the left. • Work is correctly organized, with all criteria. • The use of color and labeled diagrams enhances understanding. • The notebook is very neat.
9	**GREAT** • The writing follows the basic requirements and shows understanding of concepts, but does not go beyond. • The work shows in-depth reflection. • The notebook has all the components expected, including dates and labels on each page. • All pages are numbered properly, with odd numbers on the right and even numbers on the left. • Work is correctly organized, with all criteria. • The notebook has color and uses labeled diagrams. • The notebook looks much like one that receives a score of 10, but it lacks the perfection.
8	**GOOD** • The written work shows a basic understanding of concepts. • The work is an honest reflection, but it is limited. • The notebook has about 90% of the components expected, with dates and labels. • All pages are numbered properly, with odd numbers on the right and even numbers on the left. • Work is correctly organized. • The notebook has some color and diagrams, with a few labels. • Some requirements are met, but the notebook does not meet all criteria.
7	**FAIR** • The written work shows a limited understanding of concepts. • The work shows limited reflection overall. • The notebook has about 80% of the components, with dates and labels. • Most pages are numbered. • Work is fairly organized but just so-so. • The notebook has very little color and few diagrams. • Notebook requirements are rarely met.

Continued

STEM Research Notebook Rubric (*continued*)

6	NEEDS IMPROVEMENT
	• The written work shows misconceptions and a lack of understanding.
	• The work includes little to no reflection.
	• The pages in the notebook are unfinished.
	• There are incomplete dates and labels.
	• There are inconsistencies in right- and left-side entries.
	• The notebook is unorganized, and one or two pages are blank.
5	INCOMPLETE
	• Many pages are blank or include the class templates only.

SCORE: _____

COMMENTS:

	Rain Gauge Design Challenge Rubric (36 points possible)	
Team Name: _____		
Points Possible		*Score*
DEFINE: Identify the Problem		
5–6	• Team has a clearly stated understanding of the problem and its requirements. • Team applies knowledge to predict and solve the problem in unique and innovative ways.	
3–4	• Team can describe the problem and its requirements. • Team applies knowledge to predict and solve the problem.	
1–2	• Team needs help to understand the problem. • Team is able to recognize and mimic examples of other solutions to the problem.	
LEARN: Brainstorm Possibilities, Choose the Best		
5–6	• All ideas are focused on solving the problem. • Team considers multiple solutions and models and is able to identify those that are and are not likely to work. • Team uses ideas, concepts, or processes correctly to construct innovative possible solutions.	
3–4	• Some ideas are focused on solving the problem. • Team considers multiple solutions and models. • Team mimics (copies) the best option.	
1–2	• Few ideas are focused on solving the problem. • Team uses examples or direct guidance to generate ideas for solving the problem and developing a model. • Team does not provide any original ideas.	
PLAN: Research, List Needs, Plan Steps		
5–6	• Team critically investigates the problem and selects information from a broad range of sources. • Team provides a complete list of the materials needed. • Team plans and explains the steps that are used, including all the techniques that are required.	
3–4	• Team analyzes and selects information from some recommended sources. • Team provides a list with most of the materials needed. • Team plans and explains the steps that are used, with few omissions.	
1–2	• Team investigates the problem, collecting information from sources. • Team lists some of the materials needed. • Team plans and explains few of the steps that are needed.	

Rain Gauge Design Challenge Rubric (*continued*)

	TRY: Follow Your Plan and Build a Model	
5–6	• Team creates detailed drawings that provide valuable information for constructing the model. • Team generates a range of feasible solutions and justifies the chosen prototype.	
3–4	• Team creates drawings that provide enough information to construct a model. • Team generates an alternate solution if something doesn't work right.	
1–2	• Team provides rough sketches of the design that are not very helpful. • Team generates only one solution (even if a better alternative becomes available).	
	TEST: Try It Out! What Works? What Doesn't?	
5–6	• Team is able to manipulate materials and equipment with skill. • Team consistently adheres to plans in a precise and methodical manner. • Team does comprehensive testing to assess the performance of the model. • Test results are well documented and conclusive.	
3–4	• Team is able to manipulate materials and equipment to complete the task. • Team carries out the plan but overlooks some aspects. • The model is tested against a minimum set of standards for success. • Test results are well documented and reasonable.	
1–2	• Team is not able to manipulate equipment and materials satisfactorily. • Team has not followed the specified instructions outlined in the plan. • Team has not tested the model adequately. • Test results are inconclusive or misleading.	

Continued

Rain Gauge Design Challenge Rubric (*continued*)

DECIDE: Redesign to Solve Problems That Come Up		
5–6	• Team's evaluation is a thoughtful and insightful assessment based on the performance against a complete set of defined criteria for success. • The evaluation statements are justified and supported with evidence, taking into account the effectiveness and efficiency. • Realistic and innovative suggestions are made for improvement of the model.	
3–4	• Team's evaluations are justified and supported with evidence, taking effectiveness and efficiency of the model into consideration. • Team measures performance against most of the defined criteria for success. • Team provides some realistic suggestions to address the problem or improve performance.	
1–2	• Team states expected performance. • Team considers only one or two of the defined criteria for success for the model. • Little or no attention was given to the effectiveness or efficiency of the performance of the model.	

TOTAL SCORE: _____

COMMENTS:

Watershed Place Mat Rubric

Name: _____

Criteria	Exceeds Standard (4 points)	Meets Standard (3 points)	Approaches Standard (2 points)	Needs Improvement (1 point)	Score
DESIGN	The presentation is colorful and neat, and the information is labeled and well organized.	The presentation is neat, and the information is labeled and well organized.	The presentation is well organized with labels but is not attractive.	The presentation and organization of material are confusing to the reader.	
DETAILS	The presentation includes many interesting and clearly communicated details about the watershed or watershed conservation.	The presentation includes some interesting details that are easy to understand about the watershed or watershed conservation.	The presentation includes some details about the watershed or watershed conservation, but they may be difficult to understand.	The presentation has few details about the watershed or watershed conservation, but they are difficult to understand.	
ACCURACY	The content is accurate, and all sources are cited.	There is one error in content, and all sources are cited.	There are two to four errors in content, and all sources may not be cited.	There are more than four errors, and sources are not cited.	
GRAPHICS	The images used are excellent and help inform readers about watershed concerns.	The images used are good and adequately represent watershed concerns.	The images do not help inform the viewer, but they are related to the watershed.	Images are not used or are not related to the watershed.	
GRAMMAR AND SPELLING	There are no grammar mistakes and no spelling errors in the final project.	There are one or two grammar mistakes or spelling errors.	There are three to five grammar mistakes or spelling errors.	There are more than five grammar mistakes or spelling errors.	

TOTAL SCORE: _____

COMMENTS:

Poetry Writing Rubric

Use this rubric to assess your poem and your classmates' poems.

Name of Reader: _____

Name of Poet: _____

Criteria	Yes/No	Comments or Suggestions
IDEA/CONTENT The poem has a central idea that is clear, complete, and well developed, with supporting details.		
ORGANIZATION The structure of the poem suits the type of poetry chosen.		
WORD CHOICE The words and grammar choices express the intended message in an interesting, precise, and natural way that is appropriate for the audience and the purpose.		
FLUENCY The rhythm of the language flows, and the poem includes a variety of sentence structures.		
VOICE The writer speaks in a way that is unique, compelling, and engaging.		
GRAMMAR AND SPELLING The mechanics of writing, spelling, capitalization, and punctuation do not interfere with the reader's understanding of the poem.		
PRESENTATION The poem includes a title and the name of the author. The text is easy to read.		

Lesson Plan 2: Earth's Spheres

In this lesson, students continue to explore rainwater by examining the interconnectedness of Earth's spheres. Students investigate water handling features at home and in school while conducting surveys. They learn about the need to improvise to build tools to measure very large objects or objects in hard to reach locations. Students create biodome models and use maps to examine the idea that Earth's landforms (the geosphere) interact with the weather (the atmosphere). Students also learn how farming practices, part of the biosphere, interact with the geosphere and atmosphere. Throughout this lesson, students read and write about the spheres, gardening, water flow, and careers.

ESSENTIAL QUESTIONS

- How does precipitation interact with human-built structures?

- What information is needed to calculate the volume of rainfall on a building or surface?

- How do Earth's spheres interact?

- How does a continental divide, a special kind of land feature, provide an example of how Earth's spheres interact?

- What inferences can be made about how Earth's spheres interact by observing physical and climate maps?

ESTABLISHED GOALS AND OBJECTIVES

At the conclusion of this lesson, students will be able to do the following:

- Recognize building features that keep rainwater from entering buildings

- Use observations of rainwater handling features to predict where rainwater goes when it is channeled away from the building.

- Understand that specialized instruments are used to measure large spaces, and create and use measuring instruments to measure the school building and schoolyard

- Measure the footprint (amount of space covered by an object) of large structures such as buildings or playgrounds

- Build a scale map of a large area such as a schoolyard

- Use collaboration skills to accomplish work as members of a team

- Calculate rainfall volume on a large surface

- Identify the four Earth spheres (biosphere, geosphere, hydrosphere, and atmosphere)

- Provide examples of the ways Earth's four systems interact with each other

- Create a model showing how Earth's major spheres interact

- Use maps to make inferences about how Earth's major systems interact

- Identify the characteristics of biographical writing

- Identify the difference between primary and secondary sources

- Write a biographical text

TIME REQUIRED

- 7 days (approximately 45 minutes each day; see Tables 3.7 and 3.8, pp. 43–44)

MATERIALS

Necessary Materials for Lesson 2 (per class)

- STEM Research Notebooks (1 per student)

- Anchor charts for each of Earth's spheres (create or purchase)

- Colored markers: pink, purple, green, blue, and red (for teacher)

- Chart paper

- Handouts and rubrics (attached at the end of this lesson)

Additional Materials for Introductory Activity/Engagement (per student unless otherwise noted)

- 4 pastel sticky notes (blue, green, pink, and purple)

- 4 markers (blue, green, pink, and purple)

- 8 ½ × 11 inch piece of white construction paper

- 11 × 17 inch piece of white or light-colored construction paper

- Glue stick

- Sample landscape drawings

- *The Four Spheres of Earth,* by Paul Larson (Teacher Created Materials, 2015; for the class)

- *A Drop Around the World*, by Barbara Shaw McKinney (Dawn Publications, 1998; for the class)

- Rainwater at Home Survey handout (1 per student; see p. 137)

Additional Materials for Mathematics Class (per team unless otherwise noted)

- Padlet, Google Docs, or digital whiteboard (for the class)

- Phone or digital camera (for each student)

- 2 rulers with holes near one end

- #6 stove bolt, washer, and nut

- Small protractor

- Masking or duct tape

- 12 m heavy string or cord

- 2 dowel rods (8 inch) or unsharpened pencils

- Permanent marker

- 4 tent stakes

- Mallet or hammer

- 5 pieces colored construction paper (8½ × 11 inch)

- 1 sheet poster board

- Fine-point black marker

- Indirectly vented chemical splash goggles (1 per student)

- Surveyor Tools handout (p. 138)

- Schoolyard Surveyors handout (p. 139)

- Schoolyard Map Checklist handout (p. 144)

Additional Materials for Science Connection (per team unless otherwise noted)

- Masking tape or duct tape

- 2 plastic 2 liter bottles

- Several types of plants (see suggestions in Teacher Background Information section on p. 114)

- Small pebbles or gravel

- 2–3 cups of potting soil

- 2–3 cups of distilled water for "pond"

- 12 inch piece of cotton string

- Misting spray bottle filled with water

- Miscellaneous natural materials such as sand, pebbles or gravel, and sticks

- Insects, snails, and worms (purchase a small supply from a pet store or bait shop)

- Indirectly vented chemical splash goggles (1 per student)

- Build Your Own Biodome handout (p. 141), EDP Applied to the Biodome handout (p. 142), or both

Additional Materials for ELA Connection

- *Rachel Carson and Her Book That Changed the World,* by Laurie Lawlor (Holiday House, 2014)

Additional Materials for Social Studies Connection

- World map or globe

- Road map

- Physical Map of North America (p. 143)

- Continental Divides of North America map (e.g., *https://commons.wikimedia.org/ wiki/File%3ANorthAmerica-WaterDivides.png*)

- Agricultural plant hardiness zones map (e.g., *www.arborday.org/media/zones.cfm*)

- Precipitation map (e.g., *http://prism.oregonstate.edu/normals*)

- U.S. climate regions map (e.g., *www.pbslearningmedia.org/resource/buac17-35-sci-ess- usclimatezones/major-us-climate-zones*)

- 6 inch piece of string

- *Weslandia,* by Paul Fleischman (Candlewick Press, 2002; optional)

- Colored pencils (for each student)

- Brochures, magazines, catalogs of local flora (for each student)

SAFETY NOTES

1. Direct teacher supervision is imperative during all aspects of this activity to make sure students follow the safety guidelines.

2. The teacher should provide instruction on proper use of hand tools before the activity to help prevent skin scrapes, punctures, cuts, or other injuries.

3. All laboratory occupants must wear indirectly vented chemical splash goggles during all phases of this inquiry activity.

4. Immediately clean up any water, soil, sand, or gravel that is spilled on the floor to avoid a slip-and-fall hazard.

5. Use caution when working with sharps, which can cut or puncture skin.

6. Wash hands with soap and water after completing the activity.

CONTENT STANDARDS AND KEY VOCABULARY

Table 4.4 lists the content standards from the *NGSS, CCSS,* and the Framework for 21st Century Learning that this lesson addresses, and Table 4.5 (p. 108) presents the key vocabulary. Vocabulary terms are provided for both teacher and student use. Teachers may choose to introduce some or all of the terms to students.

Table 4.4. Content Standards Addressed in STEM Road Map Module Lesson 2

NEXT GENERATION SCIENCE STANDARDS
PERFORMANCE EXPECTATIONS
• 5-ESS2-1. Develop a model using an example to describe ways in which the geosphere, biosphere, hydrosphere, and/or atmosphere interact.
• 5-ESS2-2. Describe and graph the amounts and percentages of water and fresh water in various reservoirs to provide evidence about the distribution of water on Earth.
• 5-ESS3-1. Obtain and combine information about ways individual communities use science ideas to protect the Earth's resources and environment.
• 5-LS2-1. Develop a model to describe the movement of matter among plants, animals, decomposers, and the environment.
SCIENCE AND ENGINEERING PRACTICES
Asking Questions and Defining Problems
• Ask questions that can be investigated and predict reasonable outcomes based on patterns such as cause and effect relationships.

Continued

Table 4.4. (*continued*)

Developing and Using Models

- Identify limitations of models.

- Collaboratively develop and/or revise a model based on evidence that shows the relationships among variables for frequent and regular occurring events.

- Develop and/or use models to describe and/or predict phenomena.

- Develop a diagram or simple physical prototype to convey a proposed object, tool, or process.

- Use a model to test cause and effect relationships or interactions concerning the functioning of a natural or designed system.

Planning and Carrying Out Investigations

- Plan and conduct an investigation collaboratively to produce data to serve as the basis for evidence, using fair tests in which variables are controlled and the number of trials considered.

- Evaluate appropriate methods and/or tools for collecting data.

- Make observations and/or measurements to produce data to serve as the basis for evidence for an explanation of a phenomenon or test a design solution.

- Make predictions about what would happen if a variable changes.

Analyzing and Interpreting Data

- Represent data in tables and/or various graphical displays (bar graphs, pictographs, and/or pie charts) to reveal patterns that indicate relationships.

- Analyze and interpret data to make sense of phenomena, using logical reasoning, mathematics, and/or computation.

Using Mathematics and Computational Thinking

- Describe, measure, estimate, and/or graph quantities (e.g., area, volume, weight, time) to address scientific and engineering questions and problems.

- Create and/or use graphs and/or charts generated from simple algorithms to compare alternative solutions to an engineering problem.

Constructing Explanations and Designing Solutions

- Construct an explanation of observed relationships (e.g., the distribution of plants in the backyard).

- Use evidence (e.g., measurements, observations, patterns) to construct or support an explanation or design a solution to a problem.

- Identify the evidence that supports particular points in an explanation.

- Apply scientific ideas to solve design problems.

Continued

Table 4.4. (*continued*)

Engaging in Argument From Evidence

- Compare and refine arguments based on an evaluation of the evidence presented.

- Respectfully provide and receive critiques from peers about a proposed procedure, explanation, or model by citing relevant evidence and posing specific questions.

- Construct and/or support an argument with evidence, data, and/or a model.

- Use data to evaluate claims about cause and effect.

- Make a claim about the merit of a solution to a problem by citing relevant evidence about how it meets the criteria and constraints of the problem.

Obtaining, Evaluating, and Communicating Information

- Obtain and combine information from books and other reliable media to explain phenomena.

- Read and comprehend grade-appropriate complex texts and/or other reliable media to summarize and obtain scientific and technical ideas and describe how they are supported by evidence.

- Communicate scientific and/or technical information orally and/or in written formats, including various forms of media as well as tables, diagrams, and charts.

DISCIPLINARY CORE IDEAS

ESS2.A: Earth Materials and Systems

- Earth's major systems are the geosphere (solid and molten rock, soil, and sediments), the hydrosphere (water and ice), the atmosphere (air), and the biosphere (living things, including humans). These systems interact in multiple ways to affect Earth's surface materials and processes. The ocean supports a variety of ecosystems and organisms, shapes landforms, and influences climate. Winds and clouds in the atmosphere interact with the landforms to determine patterns of weather.

ESS2.C: The Roles of Water in Earth's Surface Processes

- Nearly all of Earth's available water is in the ocean. Most fresh water is in glaciers or underground; only a tiny fraction is in streams, lakes, wetlands, and the atmosphere.

ESS3.C: Human Impacts on Earth Systems

- Human activities in agriculture, industry, and everyday life have had major effects on land, vegetation, streams, oceans, air, and even outer space. But individuals and communities are doing things to help protect Earth's resources and environments.

LS1.C: Organization for Matter and Energy Flow in Organisms

- Plants acquire their material for growth chiefly from air and water.

Continued

Table 4.4. (*continued*)

LS2.A: Interdependent Relationships in Ecosystems
- Decomposition eventually restores (recycles) some materials back to the soil. Organisms can survive only in environments in which their particular needs are met. A healthy ecosystem is one in which multiple species of different types are each able to meet their needs in a relatively stable web of life. Newly introduced species can damage the balance of an ecosystem.

LS2.B: Cycles of Matter and Energy Transfer in Ecosystems
- Matter cycles between air and soil and among plants, animals, and microbes as these organisms live and die. Organisms obtain gases and water from the environment and release waste matter (gas, liquid, or solid) back into the environment.

CROSSCUTTING CONCEPTS

Energy and Matter
- Energy can be transferred in various ways and between objects.
- Matter is transported into, out of, and within systems.

Scale, Proportion, and Quantity
- Standard units are used to measure and describe physical quantities such as weight and volume.

Systems and System Models
- A system can be described in terms of its components and their interactions.

COMMON CORE STATE STANDARDS FOR MATHEMATICS

MATHEMATICAL PRACTICES
- MP1. Make sense of problems and persevere in solving them.
- MP2. Reason abstractly and quantitatively.
- MP3. Construct viable arguments and critique the reasoning of others.
- MP4. Model with mathematics.
- MP5. Use appropriate tools strategically.
- MP6. Attend to precision.

MATHEMATICAL CONTENT
- 5.MD.A.1. Convert among different-sized standard measurement units within a given measurement system, and use these conversions in solving multi-step, real world problems.
- 5.MD.C.3. Recognize volume as an attribute of solid figures and understand concepts of volume measurement.

Continued

Table 4.4. (*continued*)

- 5.MD.C.3.A. A cube with side length 1 unit, called a "unit cube," is said to have "one cubic unit" of volume and can be used to measure volume.
- 5.MD.C.4. Measure volumes by counting unit cubes, using cubic cm, cubic inches, cubic feet, and improvised units.
- 5.MD.C.5. Relate volume to the operations of multiplication and addition and solve real world and mathematical problems using volume.
- 5.MD.C.5.A. Find the volume of a right rectangular prism with whole-number side lengths by packing it with unit cubes, and show that the volume is the same as would be found by multiplying the edge lengths, equivalently by multiplying the height by the area of the base. Represent threefold whole-number products as volumes, e.g., to represent the associative property of multiplication.
- 5.MD.C.5.B. Apply the formulas $V = l \times w \times h$ and $V = b \times h$ for rectangular prisms to find volumes of right rectangular prisms with whole-number edge lengths in the context of solving real world and mathematical problems.
- 5.NBT.A.3.A. Read and write decimals to thousandths using base-ten numerals and number names.
- 5.NBT.A.4. Use place value understanding to round decimals to any place.
- 5.NBT.B.5. Fluently multiply multi-digit whole numbers using the standard algorithm.
- 5.NBT.B.7. Add, subtract, multiply, and divide decimals to hundredths, using concrete models or drawings and strategies based on place value, properties of operations, and/or the relationship between addition and subtraction; relate the strategy to a written method and explain the reasoning used.

COMMON CORE STATE STANDARDS FOR ENGLISH LANGUAGE ARTS

READING STANDARDS

- RI.5.1. Quote accurately from a text when explaining what the text says explicitly and when drawing inferences from the text.
- RI.5.4. Determine the meaning of general and domain-specific words and phrases in a text relevant to a *grade 5 topic or subject area.*
- RI.5.7. Draw on information from multiple print or digital sources, demonstrating the ability to locate an answer to a question quickly or to solve a problem efficiently.
- RI.5.8. Explain how an author uses reasons and evidence to support particular points in a text, identifying which reasons and evidence support which point(s).
- RI.5.9. Integrate information from several texts on the same topic in order to write or speak about the subject knowledgeably.
- RF.5.3. Know and apply grade-level phonics and word analysis skills in decoding words.
- RF.5.4.A. Read grade-level text with purpose and understanding.
- RF.5.4.B. Read grade-level prose and poetry orally with accuracy, appropriate rate, and expression on successive readings.

Continued

Table 4.4. (*continued*)

WRITING STANDARDS

- W.5.2. Write informative/explanatory texts to examine a topic and convey ideas and information clearly.

- W.5.2.B. Develop the topic with facts, definitions, concrete details, quotations, or other information and examples related to the topic.

- W.5.2.C. Link ideas within and across categories of information using words, phrases, and clauses (e.g., *in contrast, especially*).

- W.5.2.D. Use precise language and domain-specific vocabulary to inform about or explain the topic.

- W.5.4. Produce clear and coherent writing in which the development and organization are appropriate to task, purpose, and audience.

- W.5.6. With some guidance and support from adults, use technology, including the internet, to produce and publish writing as well as to interact and collaborate with others; demonstrate sufficient command of keyboarding skills to type a minimum of two pages in a single sitting.

- W.5.7. Conduct short research projects that use several sources to build knowledge through investigation of different aspects of a topic.

- W.5.8. Recall relevant information from experiences or gather relevant information from print and digital sources; summarize or paraphrase information in notes and finished work, and provide a list of sources.

- W.5.9. Draw evidence from literary or informational texts to support analysis, reflection, and research.

SPEAKING AND LISTENING STANDARDS

- SL.5.1. Engage effectively in a range of collaborative discussions with diverse partners on *grade 5 topics and texts,* building on others' ideas and expressing their own clearly.

- SL.5.1.A. Come to discussions prepared, having read or studied required material; explicitly draw on that preparation and other information known about the topic to explore ideas under discussion.

- SL.5.1.B. Follow agreed-upon rules for discussions and carry out assigned roles.

- SL.5.1.C. Pose and respond to specific questions by making comments that contribute to the discussion and elaborate on the remarks of others.

- SL.5.1.D. Review key ideas expressed and draw conclusions in light of information and knowledge gained from the discussions.

- SL.5.4. Report on a topic or text or present an opinion, sequencing ideas logically and using appropriate facts and relevant, descriptive details to support main ideas or themes; speak clearly at an understandable pace.

- SL.5.5. Include multimedia components (e.g., graphics, sound) and visual displays in presentations when appropriate to enhance the development of main ideas or themes.

- SL.5.6. Adapt speech to a variety of contexts and tasks, using formal English when appropriate to task and situation.

Continued

Table 4.4. (*continued*)

FRAMEWORK FOR 21ST CENTURY LEARNING
• Interdisciplinary Themes: Health and Safety; Environmental Literacy; Science; Mathematics
• Learning and Innovation Skills: Creativity and Innovation; Critical Thinking and Problem Solving; Communication and Collaboration
• Information, Media, and Technology Skills: Information Literacy; Media Literacy
• Life and Career Skills: Flexibility and Adaptability; Initiative and Self-Direction; Social and Cross-Cultural Skills; Productivity and Accountability; Leadership and Responsibility

Table 4.5. Key Vocabulary for Lesson 2

Key Vocabulary	Definition
abiotic	nonliving
altitude	the height of an object or point in relation to sea level
atmosphere	the envelope of gases that surrounds Earth
biodome	a closed environment created by humans that contains plants and animals in conditions similar to nature
biography	a nonfiction story of someone's life, written by another person
biosphere	the regions of Earth's surface, atmosphere, and water that are occupied by living organisms
biotic	living
channel	a passage, trough, or tube through which a liquid flows; also, to direct a liquid through a passage, trough, or tube
climate region	the usual weather conditions of an area (temperature, air pressure, humidity, precipitation, cloudiness, and winds) throughout the year
compass rose	a picture on a map that shows directions (north, south, east, west)
continental divide	an area on a continent where one side flows into a body of water (ocean or sea) and the other side flows into a different body of water
dam	a structure built to block the flow of a river for such purposes as producing electricity and providing water for irrigation
ecosystem	a complex set of relationships among living organisms and the nonliving components of an environment
equator	an imaginary line drawn around the center of Earth equally distant from both the north and south poles

Continued

Table 4.5. (*continued*)

Key Vocabulary	Definition
feature	a part of an object or structure that is especially noticeable
footprint	the shape and size of the area an object occupies on the ground
geosphere	the solid portion of Earth; Earth's crust and upper mantle including molten rock, soil, and sediments
graphic scale	a visual representation of a map's scale that uses a line segment to represent a specific distance on a map
hydrologic cycle	see *water cycle*
hydrosphere	all of Earth's water, including surface water, groundwater, and water in the atmosphere
interconnected	having all parts working together such that no part acts alone without influencing another part
landform	a natural feature of the surface of the earth, such as mountains, plains, and valleys
latitude	a measure used to identify the position of a place in terms of its distance north or south of the equator
leeward	toward the side that is sheltered from the wind or rainfall; the side of the mountain that faces away from the prevailing winds
longitude	a measure used to identify the position of a place in terms of its distance west or east of Greenwich, England
map key or legend	Information provided on a map that explains what the symbols used on the map mean
model	a representation of something, sometimes on a smaller or larger scale to help visualize or learn about something that cannot be directly observed or experimented on
perimeter	the total distance around the outside of a two-dimensional shape
physical map	a map that shows the location of landforms in addition to country borders, major cities, and significant bodies of water
planting zone	a geographic area in which a specific type of plant life grows as a result of climatic conditions of the region
precipitation map	a type of weather informational map that graphs and shows the amount of precipitation of a state or area

Continued

Table 4.5. (*continued*)

Key Vocabulary	Definition
primary source	firsthand information about a person, event, or object, such as original historical documents, interviews, letters, speeches, and other documents that have not been changed
scale	the relationship between distances in a drawing or map and the actual distances between the objects or places represented in the drawing or map
secondary source	information about a person, event, or object created by someone who did not personally experience the person, event, or object
sphere	an area of activity or a unified domain; Earth has four, the hydro-, geo-, bio-, and atmo-
surveyor	a person whose job it is to find the shapes and sizes of pieces of land
system	a group of parts working together to perform a function
terrarium	an enclosure that is adapted or prepared for keeping plants, animals, or both under seminatural conditions for observation or study
variable	a factor or condition in a scientific experiment that can be controlled or changed
water cycle	the continuous process by which water moves from land and bodies of water to the atmosphere and back to the land and bodies of water; also known as the *hydrologic cycle*
windward	the side or direction from which the wind is blowing; the side of a mountain that faces the wind

TEACHER BACKGROUND INFORMATION
Rainwater Management and Rainwater Harvesting

In mathematics, students participate in an activity that is designed to get them thinking about how rainwater is handled when it falls in an area with buildings and pavement. Generally, rainfall is evenly distributed over small geographic areas such as neighborhoods and school campuses. A relatively small amount of rainfall translates into a large volume of water when measured across a large structure such as a building or playground. Water damage is one of the leading causes of insurance claims for structures in the United States (see the information on property damage from rainwater at *https://cultureofsafety.thesilverlining.com/maintenance/property-damage-rain-water* for more information). Rain entering structures can cause structural damage as well as damage to the contents, and therefore channeling rain away from structures is a prime concern for architects and builders. The following is a summary of features incorporated into residential structures to manage rainwater:

- Sloped roofs to shed rain and melting snow

- Shingles or tiles applied to roofs in overlapping patterns to channel water downward

- Gutters around the perimeters of roofs to collect water and prevent moisture from seeping back up under the roofs by capillary action

- Gutters hung at a slight angle so that water can drain rather than sit

- Gutters attached to downspouts to drain water down to the ground

- Splash blocks at the ends of downspouts to direct water away from the foundations and out into the yards or onto sidewalks

- Siding or panels on exterior walls, with the higher rows or panels overlapping the lower ones to keep water moving down the walls

- Small overhangs over windows to direct water away from the windows

- Window glass that is sealed with a waterproof material (glazing compound) applied at an angle at the top, bottom, and sides to direct water away from the window frames

- Doorway sills that are tilted away from the houses so that water doesn't run under the doors

- Sloped sidewalks and driveways to guide water away from the houses and out to the street

- Yards that are sloped away from the houses so that water doesn't accumulate around the foundations

- Sump pumps in basements and crawlspaces to remove groundwater that collects under the floors

- Storm sewers in streets and alleys, with large underground pipes that carry the runoff water away from the neighborhood

Structural features for rainwater management can be incorporated into systems for harvesting rainwater. There are two basic types of rainwater-harvesting systems: those that collect rain falling from buildings' roofs and those that collect rain from the ground. Rain from roofs can be directed through gutters and downspouts into storage tanks, or cisterns. Alternatively, rain can be collected on the ground, through drains or pipes, and stored in drainage ponds or in tanks beneath the ground. For more information

on rainwater harvesting, see "Residential Rainwater Harvesting" at *www.thespruce.com/residential-rainwater-harvesting-1822548*.

Tools for Measuring Large Spaces

In this lesson, students measure the school building and grounds and observe where rainwater goes when it falls on the building, paved areas, and ground. Measuring a building or playground is much different from measuring a small object in the classroom. The Surveyor Tools handout (p. 138) gives hints for creating two tools that may be helpful: a measuring cord and an exterior angle gauge. Step-by-step instructions are deliberately omitted so that students can tackle the problem like engineers. Following are some basic guidelines for the tools:

- For the measuring cord, students should attach the dowels or pencils to each end of the string to serve as handles. Students should make markings all along the string to indicate each whole meter.

- The exterior angle gauge uses the principle of opposite angles to measure the angle of the corner of a building. Students should fasten the two rulers together with the bolt, washer, and nut. Then, they should tape the protractor in place to measure the opposite angle formed when the rulers are fitted to two sides of a corner like a pair of scissors.

An alternative to physically measuring the schoolyard is to use aerial maps such as those provided by Google Earth and the United States Geographical Survey (USGS). The USGS Earth Explorer tool at *http://earthexplorer.usgs.gov* offers free high-resolution photos that can serve as a starting place for a schoolyard mapping exercise. Instructions for using this tool are available on the USGS Schoolyard Geology website at *https://education.usgs.gov/lessons/schoolyard/NationalMap.html*.

Earth's Major Systems (Spheres)

All the components on Earth can be categorized within one of four major interacting subsystems, or spheres:

- *Biosphere.* The "life sphere" consists of all living things—plants and animals great and small, including insects and viruses.

- *Geosphere.* The "ground sphere" includes the ground you stand on and the whole inside of Earth—all continents, the ocean floor, sand in the desert, mountains, liquid rock, and minerals of outer and inner core. No part of the geosphere is alive.

- *Hydrosphere.* The "water sphere" is composed of all the bodies of water on Earth: rivers, lakes, and oceans, as well as ice caps, glaciers, and groundwater. It also includes the water in the air. Because more than 70% of Earth's surface is covered with water, its influence on other systems is considerable.

- *Atmosphere.* The "air sphere" is the envelope of gases that surrounds Earth. Earth's gravity holds the gases in place even though they are always in motion and constantly changing. Nitrogen, oxygen, and argon make up 99% of our atmosphere, but there are 11 other gases as well. The atmosphere is also composed of water vapor dispersed within these gases.

These spheres interact in a variety of ways and affect Earth's surface materials and processes. Here are a few examples:

- The atmosphere brings precipitation (hydrosphere) to water plants (biosphere), which draw water (hydrosphere) and nutrients from the soil (geosphere) and release water vapor back into the atmosphere.

- Humans (biosphere) use farm machinery to plow fields (geosphere).

- Humans (biosphere) build a dam out of rock materials (geosphere), and water in the lake (hydrosphere) seeps into the walls of rock (geosphere) behind the dam, becoming groundwater (hydrosphere) or evaporating into the air (atmosphere).

The balance among these complex interactions makes it possible for life on Earth to flourish. Because the spheres are all part of one interconnected system, changes in any of the spheres affect all the other spheres as well.

For more information on the differences and interactions among spheres, see the following videos:

- Four Spheres, Part 1 (Geo and Bio): Crash Course Kids #6.1, *www.youtube.com/watch?v=VMxjzWHbyFM*

- Four Spheres, Part 2 (Hydro and Atmo): Crash Course Kids #6.2, *www.youtube.com/watch?v=UXh_7wbnS3A*

Biodomes

Biodomes are ecosystems made by humans to mimic the conditions of natural ecosystems. Students will create biodome models (terrariums) in this lesson. Here are some pointers about the plants you will provide to students for their biodome models:

- Plants should be species that are able to tolerate humidity, since the enclosed environment will trap moisture.

- Plants also need to be able to tolerate low and indirect light.

- Plants should be small enough that they will not touch the sides of the container.

These are some popular plant choices for terrariums:

- Moss

- Small ferns

- Baby's tears or angel's tears (*Helxine [Soleirolia] soleirolii*; an aggressive grower)

- Miniature African violet (*Saintpaulia ionantha*)

- Creeping fig (*Ficus pumila*; an aggressive grower)

- Aluminum plant (*Pilea cadierei*)

- Nerve plant (*Fittonia albivenis*)

- Coral berry (*Ardisia crenata*)

- Buddhist pine (*Podocarpus macrophyllus*)

- Lipstick plant (*Aeschynanthus radicans*)

- Coffee plant (*Coffea arabica*)

- Creeping Charlie (*Pilea nummulariifolia*)

- Dwarf boxwood (*Buxus microphylla*)

- Purple secretia (*Tradescantia sp.*; an aggressive grower)

For the classroom biodome models, potting soil should be tested for drainage before use by squeezing a handful; if the soil clumps easily, add some vermiculite to help with drainage. This can be found in garden shops. When adding plants, keep as little of the plant's soil as possible without breaking off too many roots. The small amount that is left will provide food for several years if the plants are kept small. Tell students not to press the soil down, but to just firm the material around the plant enough to stabilize the plant. They should be careful not to overplant the biodomes, leaving plenty of room for plants to grow.

The biodomes will use a wick hanging in water to provide moisture on an ongoing basis. To provide water in the biodomes initially, have students spray water lightly around the individual plants with the misting spray bottles before sealing the biodomes. Water will spread throughout the mix by itself. Err on the side of too little rather than too much water. Then, use masking tape to close the biodome securely, and place the biodome in a bright area, but not in direct sunlight. The biodomes cannot tolerate direct sunlight because heat builds up too quickly in them. Wait at least one week to see if a condensation cycle starts. If no condensation forms, try moving the biodome to a sunnier location and add a few tablespoons of water a day until it does. If too much condensation forms, open the biodome, wipe off the condensation with a paper towel, and seal the bottle again. Do not leave the biodome open to dry it, since plants that like high humidity will suffer. If excess condensation continues to form, repeat daily. When an appropriate level is achieved, seal the bottle tightly and leave it in indirect sunlight.

The biodomes may thrive for a year or more without additional water if the proper balance of water and light has been reached, provided that it is properly sealed. Students will need to open them only for periodic housekeeping and trimming. The biodomes should require minimal care. They can be sprinkled sparingly with water when the soil is dry (provide water every couple of weeks at most). Be very careful not to overwater.

You should be alert for mold and decay. Look for thin, black fibers with tiny spore heads on moist dead leaves. This is often the early stages of moss, but it may also mean that the biosphere environment is too wet. White, furry mold can be rubbed with your fingers as soon as you notice it, and it often stays away after repeating this two or three times. Dead leaves can be left unless they are rotting against the side of the biodome. If a lot of leaves are dying, move the bottle to an area with more light and wipe condensation out to make it dryer. Tall containers seem less prone to mold issues.

Rachel Carson

In this lesson's ELA connection, the class will read aloud a picture book about Rachel Carson as an example of biographical text and an example of how writing can be used to inform the public about issues affecting the environment. Rachel Carson (1907–1964) was a marine scientist and nature writer. She worked for the U.S. Fish and Wildlife Service as a writer and editor and wrote several books about the natural world. In her best-known work, *Silent Spring,* she brought to light the dangers of pesticides, in particular the insecticide DDT (dichloro-diphenyl-trichloroethane). This work was inspired by a letter from a friend who noted unusual numbers of bird deaths around her home after the application of DDT to control mosquitoes. *Silent Spring* provided the impetus for grassroots environmental movements that eventually influenced the U.S. ban on the use of DDT.

COMMON MISCONCEPTIONS

Students will have various types of prior knowledge about the concepts introduced in this lesson. Table 4.6 (p. 116) outlines some common misconceptions students may have concerning these concepts. Because of the breadth of students' experiences, it is not possible to anticipate every misconception that students may bring as they approach this lesson. Incorrect or inaccurate prior understanding of concepts can influence student learning in the future, however, so it is important to be alert to misconceptions such as those presented in the table.

Table 4.6. Common Misconceptions About the Concepts in Lesson 2

Topic	Student Misconception	Explanation
Earth science	Because there is so much water on Earth, people do not need to be careful about how much water they use.	Although about 75% of Earth is covered with water, most of this water (about 98%) is salt water that cannot be used for drinking or for agriculture. Fresh water is therefore a scarce resource.
	Earth's spheres operate independently; for example, a change in the geosphere cannot cause a change in the atmosphere.	Earth's spheres interact in complex ways, and events in one sphere affect other spheres (e.g., a volcanic eruption in the geosphere can cause changes in the atmosphere from the release of heat and ash into the air).
	Puddles are caused by rainwater falling unevenly on surfaces.	Puddles are caused by uneven ground surfaces that result in rainwater pooling in low spots.

PREPARATION FOR LESSON 2

Review the Teacher Background Information (p. 110), assemble the materials for the lesson, and preview the videos recommended in the Learning Components section below.

To prepare for the Introductory Activity/Engagement, draw a labeled four-section square on the board representing Earth's four spheres—biosphere, geosphere, hydrosphere, and atmosphere—with space for students to post sticky notes. Create or purchase anchor charts for Earth's spheres.

Prepare for students to go outside in mathematics class to take measurements of the schoolyard and school building. Check the school grounds in advance for hazards students might encounter. You may wish to ask parents or caregivers to volunteer to assist with this activity. Students will create scaled drawings in this lesson. If students are unfamiliar with the concept of scale and how to create scaled drawings, conduct a minilesson about creating scaled drawings before the Schoolyard Surveyors activity.

For the science connection, decide whether you want to provide or have students bring clean 2 liter soda bottles (2 per student). Cut the bottles and drill holes in advance (see the Build Your Own Biodome Teacher Preparation Instructions on p. 140). Determine whether you wish to have students create their biodomes using the EDP Applied to the Biodome handout (p. 142) or following the structured instructions for the activity on the Build Your Own Biodome student handout (p. 141). Choose several types of plants that students may include in their biodomes (see the list of suggested plants in the Teacher Background Information section on p. 114). If your school policies allow, you might also include small animals such as insects, worms, or snails, but be sure to select animals that will not eat each other or endanger the plants growing in the biodome. Ensure that the animals will have a food source available within the biodome. Prepare a list of these plants and animals for students to choose from, and have the plants and animals on hand for students to place inside their biodomes.

Prepare an assortment of images to use to review Earth's spheres and their interactions with the class. For example, you may show students images of a tropical island, a forest fire, oil wells burning, and a wind farm.

For ELA, visit your local library to compile a collection of fiction and nonfiction books about Earth's spheres, environmental issues, and the water cycle. A list of suggested books for this lesson is provided on page 133. The librarian may put together a collection for you if you call ahead. Also in ELA, students will investigate biographies. Arrange for a trip to the school media center or, alternatively, secure a selection of biographies from the library, including diverse figures such as people from the entertainment industry, historical figures, athletes, people who led social movements, politicians, and scientists.

For the social studies connection, contact a local landscape or gardening business to ask for donations of plant catalogs. If these are not available, identify a website that students can use to find vegetation and garden plants suitable to your geographic region.

LEARNING COMPONENTS
Introductory Activity/Engagement

Connection to the Challenge: Begin each day of this lesson by directing students' attention to the driving question for the module: How can we use what we know about rainfall to design a system to provide water to a garden? Hold a brief discussion of student prior knowledge on this topic, creating a class list of key ideas on chart paper, or you may wish to have students create a notebook entry with this information.

Driving Question for Lesson 2: What part does water play in the interactions of Earth's four spheres, and how do the interactions among Earth's spheres affect rain?

Mathematics Class and Science Connection: Introduce the idea that Earth is a system made up of parts that work together. Have students share their ideas about what parts make up Earth (e.g., water, soil, animals and plants, air). Next, ask students where they think the water in the clouds that falls as rain comes from.

Create a chart divided into four sections, with Earth in the center (see Figure 4.2, p. 118), and label each of the four sections with one of Earth's spheres, using a different colored marker to label each sphere (biosphere = green; hydrosphere = blue; geosphere = purple; atmosphere = pink). Give each student sticky notes in the same four colors: pink, purple, blue, and green. Tell students to put their initials on the back of each one, and in

five words or less, write a description of each Earth sphere on the sticky note of the color that represents that sphere. Have students post their sticky notes on the chart. As a class, compare and contrast several student responses.

Figure 4.2. Earth's Spheres Preassessment

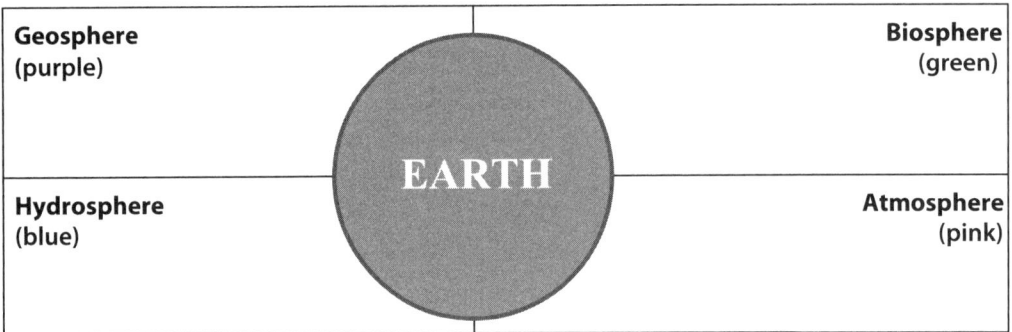

Geosphere (purple)		**Biosphere** (green)
	EARTH	
Hydrosphere (blue)		**Atmosphere** (pink)

As a class, read aloud information about Earth's spheres from students' science textbook or a book such as *The Four Spheres of Earth,* by Paul Larson. Then, hold a class discussion, asking students what information they would add to their sticky notes about each sphere. Ask students what conditions determine the way people live in an area and the type of plants and animals that live there.

Next, introduce the term *water cycle* and read aloud the book *A Drop Around the World,* by Barbara Shaw McKinney, to provide an overview of the water cycle. Hold a class discussion about what Earth spheres they observed in the book and how the hydrosphere interacted with the other spheres.

Earth's Spheres Poster

Tell students that landscape drawings and paintings include the idea that interactions between Earth's four spheres are everywhere. Pass out two pieces of construction paper (one 11 × 17 inches and one 8 ½ × 11 inches) to each student, and have students glue the 8 ½ × 11 inch piece of construction paper into the center of the 11 × 17 inch paper, leaving room along the outer edges for notes.

Project the image of one of the sample landscape drawings attached at the end of the lesson (pp. 135–136) onto the wall. Discuss the landscape image as a class, asking students to identify what spheres they see and how they think these spheres might interact. Tell students that they will create their own landscape pictures as Earth's Spheres Posters. Use the following procedure for students to create Earth's Spheres Posters:

1. Have students add the title "The Earth System" at the top of the poster. Review the definition of a system as a class.

2. Explain that Earth is a system that has four parts or subsystems, called spheres, which interact to perform functions. Review the names of the four spheres, and ask students to share what they learned about how the spheres influence each other or interact.

3. Have students create a drawing of a landscape that represents all four spheres on the 8 ½ × 11 sheet of paper. Tell students to add labels for each sphere in the margin of the paper, using a different colored marker for each sphere: biosphere = green; hydrosphere = blue; geosphere = purple; atmosphere = pink (see Figure 4.3 for an example).

Figure 4.3. The Earth System: Example Student Drawing

Note: See the book's Extras page at *www.nsta.org/roadmap-rainwater* for a larger, full-color version of this figure (the student drawing is labeled in colors and contains colored notes along the outside edge).

4. Have several students share their labeled drawings, asking the class to respond to questions about the drawings such as the following:

   • Where do you see the biosphere in this picture?

   • What are the nonliving elements of this picture?

   • What conditions do you see in the atmosphere?

   • How does the geosphere look in this picture?

   • How is the geosphere interacting with the biosphere?

   • How does the atmosphere interact with the geosphere?

5. Next, challenge students to add other elements representing at least two spheres to their pictures (for example, they may add water droplets to leaves of plants or clouds to the sky).

6. Have students work in pairs to share their pictures and their ideas about how the spheres interact in their pictures.

Emphasize to students that interactions between Earth's spheres occur all the time, and the balance among these complex interactions makes it possible for life on Earth to flourish. Ask students what part of Earth's system they are (biosphere). Ask students for their ideas about how they interact with the other spheres in their daily activities. Record student responses on chart paper. Ask students if they think these interactions are important to Earth. Tell students that although their own impact may seem small, there are over 6 billion humans on Earth, and each one affects Earth's spheres.

Introduce the idea to students that some human activities can contribute to large and visible interactions among Earth's spheres. Use as an example building dams or clearing out large areas of forests to build cities. Show a video of a large dam (for example, the CNN video "Trouble for the Three Gorges Dam" at *www.youtube.com/watch?v=YtgOntg0ofc*), asking students to observe what environmental impacts they notice and how they observe Earth's spheres interacting.

ELA Connection: Ask students about ways they can make sure that their interactions with other Earth spheres don't damage Earth. Introduce the idea that some scientists have used books to teach people about how human activity can influence other spheres. As a class, read aloud *Rachel Carson and Her Book That Changed the World,* by Laurie Lawlor. Ask students to share their ideas about how Rachel Carson changed the way humans interacted with other biospheres.

Social Studies Connection: Introduce the idea that science and engineering careers are often focused on one or more spheres. Create a T-chart that lists the names of each sphere in the left-hand column (biosphere, geosphere, hydrosphere, and atmosphere), and ask students to name careers that would relate to each sphere, writing the name of each career next to the relevant sphere. After students have provided several careers, offer additional careers (e.g., meteorologist, civil engineer, veterinarian, horticulturist, wastewater treatment plant manager) and ask students what spheres each might affect. Point out to students that many careers are related to more than one sphere, reminding students that the spheres continually interact with each other.

Activity/Exploration

Mathematics Class: Tell students that in this lesson, they will learn about how rainwater is handled when it falls in an area with buildings and pavement.

Rainwater at Home

Ask students to share their ideas about why managing rain falling on buildings might be important, recording student responses on a class list. Next, ask students to share their ideas about how they know how much rainwater falls during a rain event. Remind students that a relatively small amount of rainfall translates into a large volume of water when measured across a large structure such as a building or playground.

Ask students what the purpose of houses and other buildings are (to protect humans from weather such as cold and rain). Introduce the idea that structures that people build often include design features to prevent precipitation from entering the structure. Ask students to share their ideas about why this is important and about what these features are.

STEM Research Notebook Prompt

Have students draw a picture of a house or another building they are familiar with. Have them label the features that keep rainwater from entering the structure.

Pass out copies of the Rainwater at Home Survey handout (p. 137) for students to complete as homework. Using the handout, have students examine their homes and record observations about the buildings and how they are designed to handle rainwater. The handout includes instructions for students to sketch or photograph architectural features they observe.

STEM Research Notebook Prompt

Have students respond to the following prompt: *Describe the process you used to observe the building. Did you sketch the building or take photos? If you took photos, how do you plan to share those photos with the class? What did you observe that surprised you?*

Use the results from the Rainwater at Home Survey to discuss student observations and share their photos or sketches with the class. On the board or on chart paper, draw a large outline of a house (do not include the roof in your sketch), and have students add the rainwater management features to the house as they share their observations.

Emphasize to students that water flows down. Use the following prompt and questions to guide the discussion: Imagine that raindrops are falling from the sky and landing on your house …

- Where does the rain land first?

- What keeps the rain out of the house at that point?

- Where does the rain go next?

- What keeps the rain out of the house there?

- Where does the rain go next?

Schoolyard Surveyors

Introduce the activity with a class discussion comparing and contrasting school buildings with their homes, asking these questions:

- Does our school have features for handling rainwater similar to those on the building where you live?

What features might be found at a school that you wouldn't find at your home? Remind students that their challenge is going to be to devise ways to collect rainwater at a school, and ask students what parts of the school and schoolyard might be used to collect water.

Explain that school buildings face the same potential damage from rainwater as houses, except on a much larger scale. Point out to students that schools often have flat roofs—a special challenge when dealing with rainwater. In addition, the school grounds often have large paved areas, such as parking lots and playgrounds. In some cases, the entire school property may be paved.

Tell students that as an important first step in designing ways to collect rainwater at the school, they must find out the sizes and positions of the school building and the schoolyard. Ask students to share their ideas about why this might be important. Tell students that in this activity, they will measure the school grounds and the footprint of the school building. Explain that the footprint is the shape and size of the area an object occupies on the ground and can be represented by measuring the perimeter of the building. Students will also note observations about where rainwater goes when it falls on the building, paved areas, and the ground.

Ask students to estimate the size of the playground footprint in the schoolyard. Explain that they will obtain data about their schoolyard that they will use to calculate the total amount of water that falls on the playground during a rainstorm.

As an example, have students consider a depth dimension of ¼ inch, which for this lesson will represent approximately one hour of rainfall during a moderate rain. Make four categories on a tally chart on the board: 100–500 liters; 501–1,000 liters; 1,001–5,000 liters; >5,000 liters. Have students predict the volume in liters of water that will fall on the school building when ¼ inch of rain falls, recording tallies of student predictions on the chart.

Tell students that in order to be able to calculate the actual volume of rain that falls on the school grounds, they will measure and record the size, shape, and position of the school building and grounds, rounding measurements to the nearest meter when recording data. Ask students for their ideas about how they will take these measurements and what tools they could use, guiding students to understand that measuring large areas can be challenging using standard measuring devices such as meter sticks or tape measures. Introduce the term *surveyor* as someone who works to measure pieces of land, and introduce the idea that surveyors use special tools for this purpose. Tell students that they will act as surveyors in this activity, but that first, they will create their own tools. Distribute the Surveyor Tools (p. 138) and Schoolyard Surveyors (p. 139) handouts.

Have them work in teams to create measuring cords and exterior angle gauges, using the hints provided on the Surveyor Tools handout. (See Tools for Measuring Large Spaces on p. 112 for more information on creating these tools.)

STEM Research Notebook Prompt

Sketch the tools your team built to solve these measuring challenges. Explain how to use your tools to measure distance and angles. If you were building these tools and selling them to schools, what features could you list on an advertisement for the tools?

Students should then take measurements of their school building and schoolyard, using the following procedure along with the Schoolyard Surveyors handout. (*Note:* An option for this activity is to assign each team a different portion of the school building and schoolyard to measure.)

1. Begin at a corner and work clockwise around the object being measured.

2. Use a mallet or hammer to drive a tent stake into the ground where measurement will begin. Stretch out the measuring cord along the object.

   - If an object's side is longer than the measuring cord, place a second stake at the end of the cord, record the distance, retrieve the original tent stake, and continue measuring beginning at the second stake.

   - The length of the object is the sum of all the measurements recorded to get from one corner to the next.

3. After measuring each edge of the large object, you need to determine where the object is located so that you can create a map. Measure the distance from one corner of the object to the border of the yard that is nearest to it in each direction, as shown in Figure 4.4 on page 124 (lines *AB* and *AC*). Then, use the cord to extend the lines by sighting along the two sides that make up the corner.

Figure 4.4. Measuring the Distance From One Corner to the Nearest Border of the Yard in Each Direction

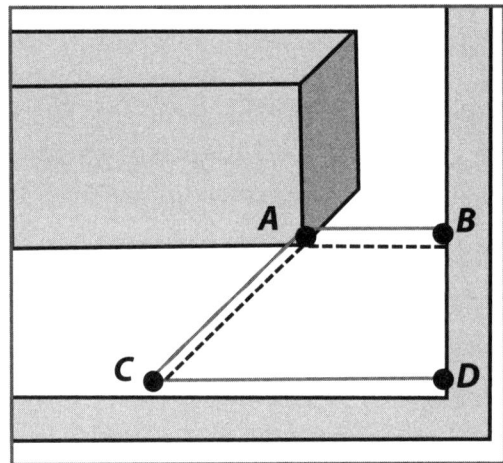

Note: Measure the distance from one corner to the nearest border of the yard in each direction by extending the lines of the intersecting sides to the borders.

4. Locate the fourth corner to make this measurement a four-sided polygon (point *D*). This may be a landmark or the edge of the yard. Measure the distance from the extended line to this point (line *CD*). By establishing these three measurements (lines *AB*, *AC*, and *CD*), you can locate the position of the object in the schoolyard.

5. Include a sketch and the dimensions in the Position of Objects section of the Schoolyard Surveyors handout.

6. Identify the surrounding streets.

7. Photograph, sketch, or describe rainwater-handling features.

8. Include notes about the features in the Water-Handling Features section of the Schoolyard Surveyors handout.

Optional: Student teams may use Google Earth or a similar resource to get dimensions, shape, and position of schoolyard objects. If they do use such resources, teams should still tour the property to gather photos and information about rainwater-handling features on the building and grounds.

STEM Research Notebook Prompt

In your STEM Research Notebook, sketch and label one of the objects you measured today. Describe a problem your team solved in order to perform this task. Describe a water-handling feature you observed while surveying the schoolyard. How does it channel water? Where does the water go after it flows through this feature?

Science Connection: Review Earth's spheres with the class. Ask students to name things that are part of the biosphere, guiding them to include plants as part of the biosphere. Show a video about plant growth that highlights what plants need to grow, such as the Veritasium video "Where Do Trees Get Their Mass From?" at *www.youtube.com/watch?v=2KZb2_vcNTg*. Hold a class discussion about plant and animal needs, asking questions such as these:

- Which spheres are addressed in this video?

- What do animals need to live and grow?

- What do plants need to live and grow?

- How do animals and plants interact?

- If plants and animals are to live together in an environment, what must be present?

Biodome Design

Introduce this activity by asking students if any of them have fish at home. Ask students where the fish live (in an aquarium). Tell students that this is an environment designed by people to mimic the conditions under which fish usually live. Tell students that they are going to create another kind of environment to mimic the living conditions of plants and small animals, called a biodome. Tell students that biodomes also provide an example of the water cycle, since the container is sealed tightly and water is recycled within it.

Divide students into their design teams, and give each team the EDP Applied to the Biodome or Build Your Own Biodome handout, whichever you decided to have students use during the lesson preparation (see p. 116). Introduce the terms *biotic* and *abiotic*. Ask students to work in their teams to brainstorm ideas about what biotic and abiotic materials they could include inside a biodome that is the size of a 2 liter bottle, prompting them to include materials from all four of Earth's spheres.

Then, have the teams share their ideas with the class. Discuss students' ideas as a class, pointing out limitations (e.g., animals such as frogs need other animals [flies] as food and require more space than a 2 liter bottle provides). Provide a list of the materials students can choose from to create their biodomes.

If you wish to take an open inquiry approach to the biodome design, using the EDP Applied to the Biodome handout, have students draw a picture of a design for their biodome using two 2 liter bottles, and answer the questions on the handout in their STEM Research Notebooks. Otherwise, have student teams follow the instructions on the Build Your Own Biodome handout to build their biodomes. (See the Teacher Background Information section on p. 113 for more details and pointers on creating and maintaining the biodomes.) Provide teams with the materials for the biodomes and have them do the following:

1. Place soil and gravel, and perhaps other natural materials, in their biodomes, according to their design.

2. Add plants in the soil of their biodomes.

3. Add animals to their biodomes.

4. Water their biodomes using the misting spray bottle before sealing it closed.

5. Seal the biodomes tightly with masking tape or duct tape so that water does not escape.

ELA Connection: Launch a class discussion by asking students to share their ideas about the differences between fiction and nonfiction literature. Create a class chart of students' responses. Ask students whether they think the book the class read earlier, *Rachel Carson and Her Book That Changed the World*, was fiction or nonfiction. Introduce the term *biography.*

Have students visit the school media center to search for biographies or, alternatively, allow them to look over a selection of biographies you have chosen in advance. Focus students' attention on the different types of people who are featured in biographies (e.g., inventors, athletes, politicians, people who led social change, entertainment industry figures). Ask students to share their ideas about what details biographies provide about people, creating a class list (e.g., where and when they were born and grew up, how they became interested in the subject or field that made them famous, what kind of education they had, what their achievements are, struggles they faced, whom they worked with). Next, reread the book *Rachel Carson and Her Book That Changed the World*, asking students to identify information from the class list that the author provided and add new items to the list.

Social Studies Connection: Ask students if the same Earth spheres are present everywhere on Earth (Yes). Ask them if the spheres look the same everywhere on Earth and how they might look different (e.g., there are oceans in some areas, lakes in others; different animals live in different parts of the world). Show students a map of North America or have them point to it on a globe. Ask students to name some of the regions of North America and describe their characteristics. (Have students consider regions rather than just states: the Midwest, Southwest, South, Northeast, the Canadian Shield, Canadian North, Baja California, Yucatan Peninsula, and so on.)

Choose two of the regions students identified, and ask students to compare the biosphere (plants and animals) that reside in the two regions, encouraging students to conclude that the location of the region influences the types of plants and animals that reside there. Ask students to share their ideas about how the physical characteristics of a region affect its biosphere (e.g., some plants thrive in the hot and dry climates found in the Southwest, animals with thick fur coats live in northern regions where it is cold).

Map Detective

Ask students to name some kinds of maps they have encountered, creating a class list of maps. Introduce the idea that there are many different types of maps, including maps that show human-created features (road maps), maps that show landforms (physical or topological maps), maps that show climate conditions, and maps that show what kinds of plants grow in a region. Show students examples of the following types of maps (see the suggested websites listed under Additional Materials for Social Studies Connection on p. 101):

- World map or globe
- Physical map of North America
- Continental Divide of North America map
- Precipitation map
- U.S. climate regions map
- Agricultural plant hardiness zones map

Using a road map as an example, introduce the terms *compass rose*, *map key or legend*, and *scale* to students. First, ask students to share their ideas about what a compass is. Ask students to identify what they see on the map that acts as a compass (the compass rose). Discuss with students the importance of knowing relative directions on a map. Next, point to the map key or legend and ask students for their ideas about the function of the key. Emphasize to students that maps can provide a variety of types of information, but because of the small space available on a map, symbols are used to identify features such as parks, different types of roads, and railroads, and that a key is necessary to understand the symbols used. Remind students that maps are representations of much larger areas. Ask students for their ideas about what would happen if they created a full size map of their classroom. Introduce the idea that maps must be small enough to be useable

and that when maps are created, each part is reduced in size by the same amount, or scaled. Point out the graphic scale on the map and, using a piece of string as a measuring device, show students how to estimate actual distances using the graphic scale.

Introduce the term *continental divide.* Have students locate each continental divide in North America as they look at the Continental Divides map and write the names of the divides in their STEM Research Notebooks.

Then, show students the Physical Map of North America (p. 143). Ask students to compare the two maps and apply context clues to describe the characteristics in the regions surrounding each continental divide. Questions that you may want students to consider include the following:

- Are the regions mostly by the coasts or inland?

- Do they mainly run in a north-south or east-west direction?

- What land formations are near most of the divides?

- What are the climate regions that surround each continental divide?

- In what direction would rainfall at each continental divide go?

- What evidence do you see on the physical map that would provide clues for what happens to rainwater as it falls?

- What spheres are interacting, and what evidence can you use to support your ideas?

Next, introduce the U.S. climate regions and precipitation maps. Hold a class discussion, asking these questions:

- How does the altitude of a region affect the atmosphere, the biosphere, the geosphere, and the hydrosphere?

- What climate similarities do states and provinces that lie along the same latitude lines have?

- What differences do states and provinces that lie along different longitudinal lines have?

- How might people (the biosphere) interact with nature in these different areas?

- How is water (the hydrosphere) represented in each of these different regions of latitude and longitude?

Encourage students to be specific in their responses, as they synthesize what they see while studying each map and gathering clues to make inferences about how the spheres interact.

Introduce the climate region and plant hardiness zones maps. Ask students to share their ideas about how these maps could be helpful when farmers and gardeners plan what to plant. Revisit the module story, Rainwater Roundup, by reminding students that Grandpa Henry and Mamito Anna live in Southern California. Tell them that the planting zone for La Vieja is zone 9. Ask students what plants Grandpa Henry and his friends might plant in their garden.

Finally, have students identify the agricultural zone that they live in, and use plant catalogs or online resources to identify the types of plants that are recommended for farmers and gardeners in this zone. Hold a class discussion, asking students to elaborate on the needs of plants that thrive in their region that are the same as or in contrast to the needs of plants in the Rainwater Roundup story. Ask students how Earth's spheres might interact to create the diverse zones.

STEM Research Notebook Prompt

Have students respond to the following prompt: *Imagine that money and space for a garden were not factors. In your STEM Research Notebook, use colored pencils to draw a plan for a community garden in your neighborhood.*

Conclude the mapping activities by reading aloud *Weslandia,* by Paul Fleischman. This story about an industrious boy provides details about the value of agriculture in our society.

Explanation

Mathematics Class: After completing their measurements in the Schoolyard Surveyors activity, each team will create a map of the schoolyard on poster board. Source data for the map should come from the Schoolyard Surveyors activity. If teams measured the entire schoolyard, each team should use its own data. If the measuring was split among teams, data sheets should be shared in either a digital or physical collaboration space.

Schoolyard Map

Provide each team with poster board, fine-point black markers, colored pencils, and the Schoolyard Map Checklist (p. 144). Review the checklist with the class. Remind students that they will use the schoolyard dimensions to consider how much rain falls on the area and how rainwater can be captured and recycled.

Science Connection: Review the four Earth spheres by providing images of various environments (e.g., a tropical island, a forest fire, oil wells burning, a wind farm) and asking students to find the interactions among the spheres.

Hold a class discussion about the biodomes student teams created, asking students to identify how the plants and animals and other features of the biodomes interact. Ask students to identify how the water cycle occurs within their biodomes.

STEM Research Notebook Prompt

Have students create sketches of their biodomes in their STEM Research Notebooks, labeling how the water cycle occurs within the biodome.

Next, have the class discuss possible outcomes for the ecosystems in the biodomes. Have students share their ideas for what will happen in a successful biodome. Students should continue making observations for the duration of this module and beyond.

ELA Connection: Ask students to share their ideas about what biographers need to do before they write a biography about someone. Guide students to understand that biographers must conduct research. Ask students how a biographer could get information about a person, creating a class list of student ideas. Introduce the terms *primary source* and *secondary source*. Ask students to review the list they created and, as a class, determine whether each item is a primary or secondary source. Ask students which source, primary or secondary, provides the best information. As a class, discuss the advantages and disadvantages of primary and secondary sources, creating a class chart of student responses. For example, primary sources may provide a great deal of detail (advantage), but a person may want to share only information that casts him or her in a positive light (disadvantage); secondary sources may be someone else's interpretation of events (disadvantage), but they can provide useful information about someone whom we cannot talk to personally (advantage).

Focus on interviews as primary sources, and ask students to list the advantages and disadvantages of using interviews to collect information for a biography. Tell students that they are each going to act as a biographer for a classmate and will conduct interviews as a source of information. Work together as a class to decide what sort of information students wish to include in the biographies. Based on the information the class decides on, have students write a list of interview questions in their STEM Research Notebooks to ask a classmate, leaving space after each question to write answers.

Classmate Biography

Pair students and have the partners interview each other, asking the questions they wrote and recording the answers. After students have completed the interviews, discuss the results, asking students what they learned about their classmates and whether there is anything more they would like to learn or anything they did not understand. Encourage students to add new questions to their lists as needed, then have students return to their pairs and ask additional questions.

Social Studies Connection: Using the climate region map and physical map, have students work as a class to identify interactions between spheres. Here are some examples:

- Mountains (geosphere) function as barriers to clouds (hydrosphere). Mountain barriers (geosphere) cause the heavy water to fall out of clouds (hydrosphere) on the windward side of the mountain (the side that gets the majority of wind and rainfall) and leaves no water for the leeward side of the mountain (the far side, which does not get the direct force of the wind and accompanying rainfall). This demonstrates how the geosphere can affect the hydrosphere, which in turn affects the geosphere.

- Sometimes a landscape is flat and barren (such as in regions around the Sahara Desert). The atmospheric winds in these areas are unobstructed and can blow in great gusts, picking up sand and debris and carrying it miles away. In these regions, the atmosphere helps shape the geosphere, but as the winds in the atmosphere crash into the hammada and oasis of the desert, they are slowed and forced to change course, and thus the atmosphere system is affected by the geosphere system.

Point out a few of the interactions that occur between the spheres globally, using context clues on the maps. Then encourage students to find and share more interactions using the maps.

Elaboration/Application of Knowledge

Mathematics Class and Science Connection: Have students work in teams to calculate the volume of rain that falls on their school building when ¼ inch of rain falls, using what they know about volume measurements. Teams should use their schoolyard maps as the source of data for the calculations. Compare students' calculations with the predictions they made at the start of the lesson. Provide scenarios with varying amounts of rainfall and varying building sizes for students to practice calculating rainfall volumes (e.g., 1 inch of rain falls on an apartment building that is 100 feet tall, 60 feet wide, and 50 feet deep).

Have student teams begin to brainstorm ways that they could collect rainwater from their school building and discuss the challenges they might face in collecting rainwater. Hold a class discussion about students' ideas.

STEM Research Notebook Prompt

Have students respond to the following prompt: *What are the challenges your team might face in collecting rainwater to reuse from your school building and grounds?*

ELA Connection: Have students use the information from their interviews to write biographies about their classmates. After students have written the biographies, have the subjects review the biographies about them and provide feedback to the authors about their accuracy.

Social Studies Connection: Not applicable.

Evaluation/Assessment

Students may be assessed on the following performance tasks and other measures listed:

Performance Tasks

- Rainwater at Home Survey
- Surveyor Tools
- Schoolyard Surveyors
- Schoolyard Map Rubric
- Earth's Spheres Poster
- Build Your Own Biodome
- EDP Applied to the Biodome (optional)
- Biography of a Classmate Rubric
- Collaboration Rubric

Other Measures

- STEM Research Notebook entries
- Engagement in class activities and discussions
- Involvement in group work and discussions

INTERNET RESOURCES

Continental Divides of North America map
- *https://commons.wikimedia.org/wiki/File%3ANorthAmerica-WaterDivides.png*

Agricultural plant hardiness zones map
- *www.arborday.org/media/zones.cfm*

Precipitation map
- *http://prism.oregonstate.edu/normals*

U.S. climate regions map
- *www.pbslearningmedia.org/resource/buac17-35-sci-ess-usclimatezones/major-us-climate-zones/#.W1-bB7hlDIU*

Property damage from rainwater
- *https://cultureofsafety.thesilverlining.com/maintenance/property-damage-rain-water*

"Residential Rainwater Harvesting"
- *www.thespruce.com/residential-rainwater-harvesting-1822548*

USGS Earth Explorer tool
- *http://earthexplorer.usgs.gov*

USGS Schoolyard Geology instructions on using the Earth Explorer tool
- *http://education.usgs.gov/lessons/schoolyard/NationalMap.html*

"Four Spheres, Part 1 (Geo and Bio): Crash Course Kids #6.1" video
- *www.youtube.com/watch?v=VMxjzWHbyFM*

"Four Spheres, Part 2 (Hydro and Atmo): Crash Course Kids #6.2" video
- *www.youtube.com/watch?v=UXh_7wbnS3A*

CNN's "Trouble for the Three Gorges Dam" video
- *www.youtube.com/watch?v=YtgOntg0ofc*

"Where Do Trees Get Their Mass From?" video
- *www.youtube.com/watch?v=2KZb2_vcNTg*

SUGGESTED BOOKS FOR LESSON 2

- *A Writing Kind of Day: Poems for Young Poets,* by Ralph Fletcher (WordSong, 2005)
- *Career Ideas for Kids Who Like Science,* by Diane Lindsey Reeves and Lindsey Clasen (Checkmark Books, 2007)
- *Cloud Dance,* by Thomas Locker (HMH Books for Young Readers, 2003)
- *Come on Rain!* by Karen Hesse (Scholastic Press, 1999)
- *Down the Drain: Conserving Water,* by Anita Ganeri and Chris Oxlade (Heinemann, 2005)

- *Exploring Ecosystems With Max Axiom, Super Scientist*, by Agnieszka Biskup (Capstone Press, 2007)

- *Hip Hop Speaks to Children: A Celebration of Poetry With a Beat*, edited by Nikki Giovanni (Sourcebooks Jabberwocky, 2008)

- *How the Earth Works*, by John Farndon (Readers Digest Press, 1992)

- *Poetry for Young People: Robert Frost*, edited by Gary D. Schmidt (Sterling, 2008)

- *The Four Spheres of Earth*, by Paul Larson (Teacher Created Materials, 2015)

- *The Hydrosphere: Agent of Change*, by Gregory Vogt (Twenty First Century Books, 2006)

- *Weslandia*, by Paul Fleischman (Candlewick Press, 2002)

- *Where the Sidewalk Ends*, by Shel Silverstein (HarperCollins, 2014)

IMAGE CREDITS FOR LESSON 2

Rain on a rooftop (p. 137)
www.flickr.com/photos/20009658@N00/24136828691
Owner: Jason Rosenberg; License: Attribution 2.0 Generic (CC BY 2.0)

Surveyors (p. 138)
www.flickr.com/photos/46645540@N00/7986956631
Owner: Sunburned Surveyor; License: Attribution 2.0 Generic (CC BY 2.0)

Drawings and charts (Owner: Pandaia Projects LLC. Used with permission.)

- Figure 4.2 (p. 118)

- Sample Earth system drawings (pp. 119, 135, and 136)

- Figure 4.4 (p. 124)

SAMPLE LANDSCAPE DRAWING #1

SAMPLE LANDSCAPE DRAWING #2

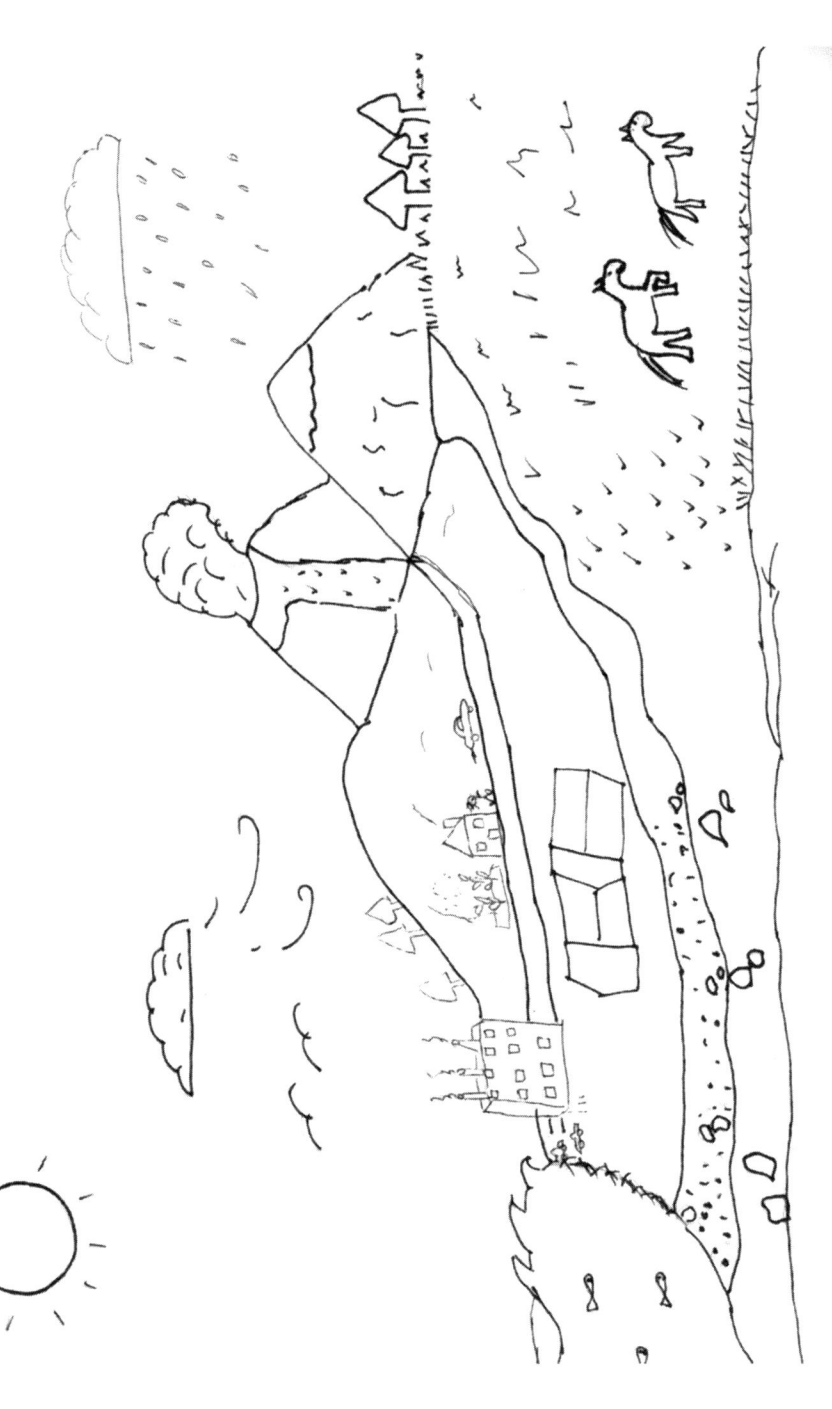

Name: _____   Date: _____

RAINWATER AT HOME SURVEY

You will make observations about how rainwater is handled around your home.

In class, we learned that water from rain and snow is important for life. But if water from rain or snow gets inside a building, such as a house, apartment building, or school, it can cause serious damage. Buildings are designed so that precipitation (rainwater and snow) is kept outside the house. For this assignment, you will look around the building where you live for structural features that keep water out. Observe the building from as many different views as possible—look out of the windows, walk around outside, look up high and down low. If possible, take photos or make sketches of the things you find to share with the class.

Describe the building where you live:

List features of the building that keep water out:

Team Name: _____   Date: _____

STUDENT HANDOUT

SURVEYOR TOOLS

Your team will measure your school building and schoolyard. This a large space, and you will need special tools to measure it.

Surveyors take measurements to find the shape and size of a piece of land. They use mathematics and special tools for their work because the objects they measure can be very large, such as a house or other building, a bridge or dam, or even a highway.

In this activity, you will solve two problems: how to measure long distances and how to measure angles of large objects.

Measuring Cord	Exterior Angle Gauge
Problem: It would be hard work to measure the length of a side of a building or playground with a ruler or meter stick.	**Problem:** To describe the shape of a large object, you need to know the angles of the sides. If the object is a building or structure, measuring the angle with a protractor will be difficult. In this case, use opposite angles to get your measurement.
Materials: You have been given a long piece of heavy string, 2 dowel rods or pencils, a permanent marker, and masking or duct tape.	**Materials:** You have been given 2 rulers; a nut, bolt, and washer; masking or duct tape; and a protractor.
Criteria: Use the string to make a measuring cord that your team can use to measure distances up to 10 meters. You should be able to read measurements on the cord and determine distances shorter than 10 meters to the nearest meter.	**Criteria:** Your exterior angle gauge should be able to measure angles to the nearest 5 degrees.

Team Name: _____   Date: _____

SCHOOLYARD SURVEYORS

Using this handout as a guide, you will do what surveyors do: provide data for the mapping team. With this information, your team will be able to create a map of the schoolyard. Take measurements of your school building and schoolyard, using the tools you made, and record your measurements in the Schoolyard Survey Data table.

While you take your measurements, look for rainwater-handling features on your school building and in the schoolyard, and take photos or make sketches of them in your STEM Research Notebooks.

SCHOOLYARD SURVEY DATA

Identify the Structure:	
Perimeter Measurements: Angle Measurements:	Sketch and Label the Structure:
Position of Objects:	Positioning Sketch:
Water-Handling Features:	

TEACHER PREPARATION INSTRUCTIONS

BUILD YOUR OWN BIODOME

1. Cut the 2 liter bottles as shown, discarding the shorter bottom piece.

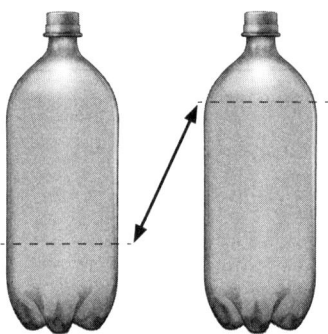

Cut along dotted lines.

2. Drill a hole through the center of one bottle cap, and thread a cotton string through it.

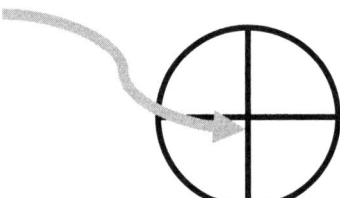

String threaded through the hole in the cap will hang down into the water in the "pond" below.

BUILD YOUR OWN BIODOME

1. The three pieces of the bottle will be connected as shown. The cap with the string threaded through the hole should be on the bottom end that rests over the base.

2. Add distilled water in the bottom to create a "pond," then position the bottle top with the hole in the cap upside down as shown, so that the string hangs down into the water to serve as a wick. Add about ½ inch of small pebbles or gravel in the upside-down bottle, then add 8 to 10 cm of potting soil.

3. Plant several different plants in the soil.

4. Add any animals (e.g., insects, snails, worms).

5. Put the other bottle top on as a "lid," and seal the biodome sections together tightly using tape.

Caps will be on both ends of the bottle on top. Top bottle sits inside the bottom base.

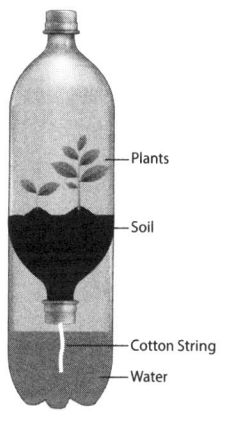

Plants

Soil

Cotton String

Water

STUDENT HANDOUT

EDP APPLIED TO THE BIODOME

DEFINE:
Identify the problem

DEFINE: WHAT PROBLEMS DO YOU WANT TO SOLVE?
1. What kind of plants do you want to have?
2. What kind of animals do you want to have?
3. How can you find out if they can live together?
4. What nonliving (abiotic) items will be in your ecosystem?

LEARN:
Brainstorm

PLAN:
Research and sketch

LEARN AND PLAN!
5. What materials will you need? Be specific.
6. What are the needs of the plants you plan to put in the biodome?
7. What are the needs of the animals you plan to put in the biodome?
8. How will you make sure these needs are going to be met?
9. Sketch the design for your biodome.
10. Label each part and show where your plants and animals will live.
11. Label where you will put the abiotic parts of your habitat.

TRY:
Build model with plan

TEST:
What works and what doesn't?

TRY AND TEST AND TRY AND TEST AND ...
12. Build your biodome! You may not be able to do this all at once; that is OK. Your teacher can give you masking tape to use to seal it each day until you have it completed.
13. Be sure to follow your plan exactly! You can make improvements later.

DECIDE:
Redesign to solve problems

DECIDE
14. Describe some observations you are making about how the environment and organisms are interacting.
15. Explain how this shows how Earth's spheres interact.
16. Are your plants and animals thriving? Why or why not?
17. What could you change to make it better?

SHARE:
Get feedback

SHARE AND GET FEEDBACK
18. Share with the class. Tell your class about the experience. What went really well? What surprised you the most? What was the biggest challenge?
19. Ask the class to give you suggestions for making it better.
20. Now is the time to finalize your design. You are able to make one change to your biodome. What will this change be?

4

PHYSICAL MAP OF NORTH AMERICA

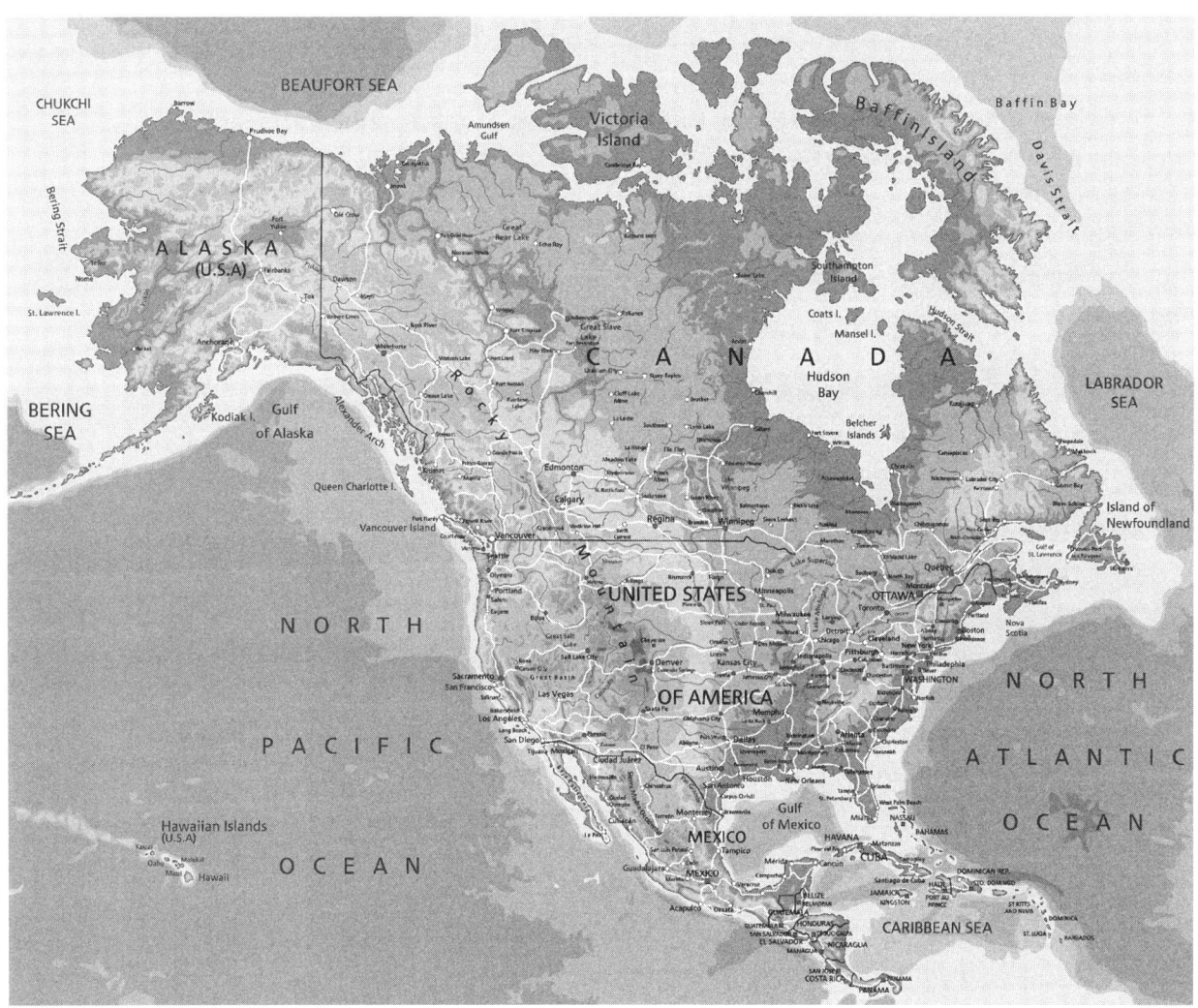

Note: A full-color version of this image is available on the book's Extras page at *www.nsta.org/roadmap-rainwater*. Colors on the full-color map indicate altitude, not conditions on the ground. Dark brown = highest altitude; dark green = near sea level; light blue = continental shelf; dark blue = deep ocean.

STUDENT HANDOUT

SCHOOLYARD MAP CHECKLIST

A map is a picture of a place. Different types of maps show different information. Draw a clear and precise map of your schoolyard on poster board.

REQUIRED FEATURES

- ✓ The map has the name of your school in the title.
- ✓ The map shows the large objects (e.g., buildings, playgrounds, parking areas) of the school campus.
- ✓ The large objects are labeled appropriately, with the names spelled correctly.
- ✓ Each object has the dimensions shown for its sides, and its area is calculated in square meters.
- ✓ Large objects are positioned correctly using construction paper footprints.
- ✓ Water-handling features have been noted. (These do not need to be to scale.)
- ✓ A compass rose shows the directions on the map.
- ✓ A scale shows distance equivalents.
- ✓ A key or legend is used to explain the symbols used on the map (such as downspout, drain, ditch).
- ✓ The map is labeled and outlined in black marker.

Schoolyard Map Rubric

Team Name: _____

Criteria	Exceeds Expectations (4 points)	Meets Expectations (3 points)	Approaches Expectations (2 points)	Needs Improvement (1 point)	Score
TITLE	The title tells the purpose/content of the map, is located in clear view, and shows extra attention to detail in its creativity.	The title tells the purpose/content of the map and is located in clear view.	The title tells the purpose/content of the map but is not in clear view.	The title is missing or does not reflect the purpose/content of the map.	
LARGE OBJECTS	All buildings, playgrounds, and paved areas are present and labeled.	Most buildings, playgrounds, and paved areas are present and labeled.	Some buildings, playgrounds, and paved areas are present and labeled.	Some buildings, playgrounds, and paved areas are present but are not labeled.	
WATER-HANDLING FEATURES	Sketches or photos of water-handling features are included for each object.	Sketches or photos of water-handling features are included for some objects.	Few sketches or photos of water-handling features are included.	No sketches or photos of water-handling features are included.	
NEATNESS	All the objects and features are clearly labeled and easy to read.	Most of the objects and features are clearly labeled and easy to read.	Some of the objects and features are clearly labeled and easy to read.	Few of the objects and features are clearly labeled and may be difficult to read.	
ACCURACY	All the objects and features are correctly placed on the map.	Most of the features are correctly placed on the map.	Some of the features are correctly placed on the map.	Few of the features are correctly placed on the map.	
SPELLING AND CAPITALIZATION	All words on the map are spelled and capitalized correctly.	Most words on the map are spelled and capitalized correctly.	Many words on the map are not spelled correctly; all words are capitalized.	Most words are not spelled correctly or capitalized.	

TOTAL SCORE: _____

COMMENTS:

Biography of a Classmate Rubric

Use this rubric to assess the biography your classmate wrote about you.

Name of Reader: _____   Name of Biographer: _____

Criteria	Yes/No	Comments or Suggestions
CONTENT The content is accurate and includes well-developed supporting details.		
ORGANIZATION The structure of the biography is clear; it has a beginning, middle, and end.		
WORD CHOICE The words and grammar used express the intended message in an interesting and descriptive way.		
FLUENCY The rhythm of the language flows, and the biography includes a variety of sentence structures.		
VOICE The writer speaks in a way that is unique, compelling, and engaging.		
GRAMMAR AND SPELLING The mechanics of writing, spelling, capitalization, and punctuation do not interfere with the reader's understanding.		
PRESENTATION The biography has a title that includes the subject's name, and the name of the author is present. The text is easy to read.		

Collaboration Rubric

Name: _____   Topic: _____

Expectations	Genius (4 points)	Expert (3 points)	Apprentice (2 points)	Developing (1 point)	Keep Trying (0 points)
SELF-REFLECTION					
I stayed on task while working with my team.					
I communicated with my team and reported our ideas and work accurately.					
I read and followed the directions carefully.					
We were successful and completed the activity, and I did my best.					
I did the same amount of work as the others on my team.					

COMMENTS:

Continued

Collaboration Rubric (*continued*)

TEACHER INPUT

Expectations	Genius (4 points)	Expert (3 points)	Apprentice (2 points)	Developing (1 point)	Keep Trying (0 points)
Student had an equal share of the responsibility and did not let his or her team down.					
Student's understanding of the problem benefited the team and contributed to the team's success.					
Student's interactions with the group were respectful and courteous.					
Student stayed on task and followed directions.					

COMMENTS:

Continued

Collaboration Rubric (*continued*)

PEER EVALUATION

Expectations	Genius (4 points)	Expert (3 points)	Apprentice (2 points)	Developing (1 point)	Keep Trying (0 points)
This team member shared responsibility for our team's work.					
This team member made good decisions that helped our team stay on task and do our best.					
This team member used his or her strengths to contribute to the success of our team.					
This team member maintained a positive attitude about the project.					

COMMENTS:

Lesson Plan 3: How Much Rain Can We Catch?

In this lesson, students continue to address the Rainwater Roundup Challenge as they determine the best shape and dimensions for a storage container for rainwater collection. Students compare their rain gauge data with rainfall data from previous years and use graphing to track monthly rain totals using historical data. Students also research strengths and weaknesses of irrigation and water collection systems.

ESSENTIAL QUESTIONS

- How does our rain gauge data compare with rain data from the past?

- How can we determine the optimal size of a rainwater storage container?

- What are some of the effects of agriculture on the environment?

- How can farmers and other people in agriculture protect groundwater and other elements of the hydrosphere?

- How can we make others aware of ways to care for Earth's hydrosphere?

ESTABLISHED GOALS AND OBJECTIVES

At the conclusion of this lesson, students will be able to do the following:

- Chart rain totals from their rain gauges to determine the amount of rainfall

- Calculate rainwater volume on various parts of the schoolyard

- Collect and graph historical rainfall data

- Create and analyze a graph to compare and contrast schoolyard rainfall measurements with statistics for rainfall in the surrounding area in a typical year

- Design a tank to hold a given amount of water.

- Describe capillary action and identify it as the mechanism by which water is moved from soil to various parts of plants

- Provide examples of ways that human activities (the biosphere) can have negative impacts on Earth's other spheres

- Propose some ways that humans can protect the hydrosphere

- Identify and describe various irrigation techniques

- Evaluate agricultural practices and propose alternative strategies for those that may have harmful impacts on the environment

- Construct a message about watershed protection using persuasive writing techniques

TIME REQUIRED

- 5 days (approximately 45 minutes each day; see Tables 3.8 and 3.9, pp. 44–45)

MATERIALS

Necessary Materials for Lesson 3

- STEM Research Notebooks
- Schoolyard Map and Schoolyard Surveyors data from Lesson 2
- Internet access
- Handouts and rubrics (attached at the end of this lesson)

Additional Materials for Mathematics Class (per team)

- Local precipitation data for past 12 months
- Rain gauge data from current month
- Computer spreadsheet application or paper spreadsheet
- Computer graphing application or paper equivalent
- Piece of construction paper
- School glue or glue stick
- Notecard
- What's the Volume? handout (p. 178)
- How Big Is Big Enough? handout (p. 180)
- Using the NOAA Website handout (p. 182)

Additional Materials for Social Studies Connection (per class)

- Sheet of poster paper
- Colored markers
- Chart paper

Additional Materials for Science Connection (per team unless otherwise noted)

- 200 or 250 ml glass beaker or a similar-size clear plastic cup

- 8 or 9 oz. water

- Red or blue food coloring

- 4 ribs of celery with leaves, 5–6 inches long

- Plastic knife

- Ruler

- Colored pencils or markers

- 2 clear plastic 10 oz. cups

- 4 paper towels

- Permanent marker

- Way to Flow: Capillary Action handout (1 per student; see p. 185)

- The Great Escape: Capillary Action handout (1 per student; see p. 186)

Additional Materials for ELA Connection

- Writing the OREO Way handout (1 per student and 1 per team; see p. 187)

SAFETY NOTES

1. Direct teacher supervision is imperative during all aspects of this activity to make sure students follow the safety guidelines.

2. All laboratory occupants must wear indirectly vented chemical splash goggles during all phases of this inquiry activity.

3. Immediately wipe up any water that is spilled on the floor to avoid a slip-and-fall hazard.

4. Use caution when working with sharps, which can cut or puncture skin.

5. Handle glassware and plasticware with care, as these can break and cut or puncture skin.

6. Wash hands with soap and water after completing the activity.

CONTENT STANDARDS AND KEY VOCABULARY

Table 4.7 lists the content standards from the *NGSS, CCSS,* and the Framework for 21st Century Learning that this lesson addresses, and Table 4.8 (p. 159) presents the key vocabulary. Vocabulary terms are provided for both teacher and student use. Teachers may choose to introduce some or all of the terms to students.

Table 4.7. Content Standards Addressed in STEM Road Map Module Lesson 3

NEXT GENERATION SCIENCE STANDARDS

PERFORMANCE EXPECTATIONS

- 5-ESS2-1. Develop a model using an example to describe ways in which the geosphere, biosphere, hydrosphere, and/or atmosphere interact.

- 5-ESS2-2. Describe and graph the amounts and percentages of water and fresh water in various reservoirs to provide evidence about the distribution of water on Earth.

- 5-ESS3-1. Obtain and combine information about ways individual communities use science ideas to protect the Earth's resources and environment.

- 5-LS2-1. Develop a model to describe the movement of matter among plants, animals, decomposers, and the environment.

SCIENCE AND ENGINEERING PRACTICES

Asking Questions and Defining Problems

- Ask questions that can be investigated and predict reasonable outcomes based on patterns such as cause and effect relationships.

Developing and Using Models

- Identify limitations of models.

- Collaboratively develop and/or revise a model based on evidence that shows the relationships among variables for frequent and regular occurring events.

- Develop and/or use models to describe and/or predict phenomena.

- Develop a diagram or simple physical prototype to convey a proposed object, tool, or process.

- Use a model to test cause and effect relationships or interactions concerning the functioning of a natural or designed system.

Planning and Carrying Out Investigations

- Plan and conduct an investigation collaboratively to produce data to serve as the basis for evidence, using fair tests in which variables are controlled and the number of trials considered.

- Evaluate appropriate methods and/or tools for collecting data.

Continued

Table 4.7. (*continued*)

- Make observations and/or measurements to produce data to serve as the basis for evidence for an explanation of a phenomenon or test a design solution.
- Make predictions about what would happen if a variable changes.

Analyzing and Interpreting Data

- Represent data in tables and/or various graphical displays (bar graphs, pictographs, and/or pie charts) to reveal patterns that indicate relationships.
- Analyze and interpret data to make sense of phenomena, using logical reasoning, mathematics, and/or computation.
- Compare and contrast data collected by different groups in order to discuss similarities and differences in their findings.
- Analyze data to refine a problem statement or the design of a proposed object, tool, or process.
- Use data to evaluate and refine design solutions.

Using Mathematics and Computational Thinking

- Describe, measure, estimate, and/or graph quantities (e.g., area, volume, weight, time) to address scientific and engineering questions and problems.
- Create and/or use graphs and/or charts generated from simple algorithms to compare alternative solutions to an engineering problem.

Constructing Explanations and Designing Solutions

- Construct an explanation of observed relationships (e.g., the distribution of plants in the backyard).
- Use evidence (e.g., measurements, observations, patterns) to construct or support an explanation or design a solution to a problem.
- Identify the evidence that supports particular points in an explanation.
- Apply scientific ideas to solve design problems.
- Generate and compare multiple solutions to a problem based on how well they meet the criteria and constraints of the design.

Engaging in Argument From Evidence

- Compare and refine arguments based on an evaluation of the evidence presented.
- Respectfully provide and receive critiques from peers about a proposed procedure, explanation, or model by citing relevant evidence and posing specific questions.
- Construct and/or support an argument with evidence, data, and/or a model.
- Use data to evaluate claims about cause and effect.
- Make a claim about the merit of a solution to a problem by citing relevant evidence about how it meets the criteria and constraints of the problem.

Continued

Table 4.7. (*continued*)

Obtaining, Evaluating, and Communicating Information

- Obtain and combine information from books and other reliable media to explain phenomena.

- Read and comprehend grade-appropriate complex texts and/or other reliable media to summarize and obtain scientific and technical ideas and describe how they are supported by evidence.

- Combine information in written text with that contained in corresponding tables, diagrams, and/or charts to support the engagement in other scientific and/or engineering practices.

- Communicate scientific and/or technical information orally and/or in written formats, including various forms of media as well as tables, diagrams, and charts.

DISCIPLINARY CORE IDEAS

ESS2.A: Earth Materials and Systems

- Earth's major systems are the geosphere (solid and molten rock, soil, and sediments), the hydrosphere (water and ice), the atmosphere (air), and the biosphere (living things, including humans). These systems interact in multiple ways to affect Earth's surface materials and processes. The ocean supports a variety of ecosystems and organisms, shapes landforms, and influences climate. Winds and clouds in the atmosphere interact with the landforms to determine patterns of weather.

ESS2.C: The Roles of Water in Earth's Surface Processes

- Nearly all of Earth's available water is in the ocean. Most fresh water is in glaciers or underground; only a tiny fraction is in streams, lakes, wetlands, and the atmosphere.

ESS3.C: Human Impacts on Earth Systems

- Human activities in agriculture, industry, and everyday life have had major effects on land, vegetation, streams, oceans, air, and even outer space. But individuals and communities are doing things to help protect Earth's resources and environments.

LS2.A: Interdependent Relationships in Ecosystems

- Decomposition eventually restores (recycles) some materials back to the soil. Organisms can survive only in environments in which their particular needs are met. A healthy ecosystem is one in which multiple species of different types are each able to meet their needs in a relatively stable web of life. Newly introduced species can damage the balance of an ecosystem.

LS2.B: Cycles of Matter and Energy Transfer in Ecosystems

- Matter cycles between air and soil and among plants, animals, and microbes as these organisms live and die. Organisms obtain gases and water from the environment and release waste matter (gas, liquid, or solid) back into the environment.

Continued

Table 4.7. (*continued*)

CROSSCUTTING CONCEPTS

Energy and Matter
- Energy can be transferred in various ways and between objects.
- Matter is transported into, out of, and within systems.

Scale, Proportion, and Quantity
- Standard units are used to measure and describe physical quantities such as weight and volume.

Systems and System Models
- A system can be described in terms of its components and their interactions.

COMMON CORE STATE STANDARDS FOR MATHEMATICS

MATHEMATICAL PRACTICES
- MP1. Make sense of problems and persevere in solving them.
- MP2. Reason abstractly and quantitatively.
- MP3. Construct viable arguments and critique the reasoning of others.
- MP4. Model with mathematics.
- MP5. Use appropriate tools strategically.
- MP6. Attend to precision.
- MP7. Look for and make use of structure.

MATHEMATICAL CONTENT
- 5.GA.1. Use a pair of perpendicular number lines, called axes, to define a coordinate system, with the intersection of lines (the origin) arranged to coincide with the 0 on each line and a given point in the plane located by using an ordered pair of numbers, called its coordinates. Understand that the first number indicates how far to travel from the origin in the direction of one axis, and the second number indicates how far to travel in the direction of the second axis, with the convention that the names of the two axes and the coordinates correspond (e.g., *x*-axis and *x*-coordinate, *y*-axis and *y*-coordinate).
- 5.GA.2. Represent real world and mathematical problems by graphing points in the first quadrant of the coordinate plane, and interpret coordinate values of points in the context of the situation.
- 5.MD.A.1. Convert among different-sized standard measurement units within a given measurement system, and use these conversions in solving multi-step, real world problems.
- 5.MD.C.3. Recognize volume as an attribute of solid figures and understand concepts of volume measurement.

Continued

Table 4.7. (*continued*)

- 5.MD.C.3.A. A cube with side length 1 unit, called a "unit cube," is said to have "one cubic unit" of volume and can be used to measure volume.

- 5.MD.C.4. Measure volumes by counting unit cubes, using cubic cm, cubic inches, cubic feet, and improvised units.

- 5.MD.C.5. Relate volume to the operations of multiplication and addition and solve real world and mathematical problems using volume.

- 5.MD.C.5.B. Apply the formulas $V = l \times w \times h$ and $V = b \times h$ for rectangular prisms to find volumes of right rectangular prisms with whole-number edge lengths in the context of solving real world and mathematical problems.

- 5.NBT.A.3.A. Read and write decimals to thousandths using base-ten numerals and number names.

- 5.NBT.A.4. Use place value understanding to round decimals to any place.

- 5.NBT.B.5. Fluently multiply multi-digit whole numbers using the standard algorithm.

- 5.NBT.B.7. Add, subtract, multiply, and divide decimals to hundredths, using concrete models or drawings and strategies based on place value, properties of operations, and/or the relationship between addition and subtraction; relate the strategy to a written method and explain the reasoning used.

COMMON CORE STATE STANDARDS FOR ENGLISH LANGUAGE ARTS

READING STANDARDS

- RI.5.1. Quote accurately from a text when explaining what the text says explicitly and when drawing inferences from the text.

- RI.5.4. Determine the meaning of general and domain-specific words and phrases in a text relevant to a *grade 5 topic or subject area.*

- RI.5.7. Draw on information from multiple print or digital sources, demonstrating the ability to locate an answer to a question quickly or to solve a problem efficiently.

- RI.5.8. Explain how an author uses reasons and evidence to support particular points in a text, identifying which reasons and evidence support which point(s).

- RI.5.9. Integrate information from several texts on the same topic in order to write or speak about the subject knowledgeably.

- RF.5.3. Know and apply grade-level phonics and word analysis skills in decoding words.

- RF.5.4.A. Read grade-level text with purpose and understanding.

- RF.5.4.B. Read grade-level prose and poetry orally with accuracy, appropriate rate, and expression on successive readings.

Continued

Table 4.7. (*continued*)

WRITING STANDARDS

- W.5.1. Write opinion pieces on topics or texts, supporting a point-of-view with reasons and information.

- W.5.2. Write informative/explanatory texts to examine a topic and convey ideas and information clearly.

- W.5.2.B. Develop the topic with facts, definitions, concrete details, quotations, or other information and examples related to the topic.

- W.5.2.C. Link ideas within and across categories of information using words, phrases, and clauses (e.g., *in contrast, especially*).

- W.5.2.D. Use precise language and domain-specific vocabulary to inform about or explain the topic.

- W.5.2.E. Provide a concluding statement or section related to the information or explanation presented.

- W.5.4. Produce clear and coherent writing in which the development and organization are appropriate to task, purpose, and audience.

- W.5.5. With guidance and support from peers and adults, develop and strengthen writing as needed by planning, revising, editing, rewriting, or trying a new approach.

- W.5.6. With some guidance and support from adults, use technology, including the internet, to produce and publish writing as well as to interact and collaborate with others; demonstrate sufficient command of keyboarding skills to type a minimum of two pages in a single sitting.

- W.5.7. Conduct short research projects that use several sources to build knowledge through investigation of different aspects of a topic.

- W.5.8. Recall relevant information from experiences or gather relevant information from print and digital sources; summarize or paraphrase information in notes and finished work, and provide a list of sources.

- W.5.9. Draw evidence from literary or informational texts to support analysis, reflection, and research.

SPEAKING AND LISTENING STANDARDS

- SL.5.1. Engage effectively in a range of collaborative discussions with diverse partners on *grade 5 topics and texts*, building on others' ideas and expressing their own clearly.

- SL.5.1.A. Come to discussions prepared, having read or studied required material; explicitly draw on that preparation and other information known about the topic to explore ideas under discussion.

- SL.5.1.B. Follow agreed-upon rules for discussions and carry out assigned roles.

- SL.5.1.C. Pose and respond to specific questions by making comments that contribute to the discussion and elaborate on the remarks of others.

Continued

Table 4.7. (continued)

- SL.5.1.D. Review key ideas expressed and draw conclusions in light of information and knowledge gained from the discussions.

- SL.5.4. Report on a topic or text or present an opinion, sequencing ideas logically and using appropriate facts and relevant, descriptive details to support main ideas or themes; speak clearly at an understandable pace.

- SL.5.5. Include multimedia components (e.g., graphics, sound) and visual displays in presentations when appropriate to enhance the development of main ideas or themes.

- SL.5.6. Adapt speech to a variety of contexts and tasks, using formal English when appropriate to task and situation.

FRAMEWORK FOR 21ST CENTURY LEARNING

- Interdisciplinary Themes: Health and Safety; Environmental Literacy; Science; Mathematics

- Learning and Innovation Skills: Creativity and Innovation; Critical Thinking and Problem Solving; Communication and Collaboration

- Information, Media, and Technology Skills: Information Literacy; Media Literacy

- Life and Career Skills: Flexibility and Adaptability; Initiative and Self-Direction; Social and Cross-Cultural Skills; Productivity and Accountability; Leadership and Responsibility

Table 4.8. Key Vocabulary for Lesson 3

Key Vocabulary	Definition
adhesion	the action or process of attaching to something even if composed of different substances
algae bloom	a rapid increase or growth in the population of microscopic plants in a water system
book series	a sequence of books that have common characteristics that identify them as a group
campaign	a series of actions meant to achieve a specific goal or result
capillary action	the ability of a liquid to flow within the spaces of porous materials, making it appear to defy the law of gravity
climate	the average weather common to a place over several years
cohesion	the tendency of particles of the same material to stick together
dimension	the length, width, or height of something
fertilizer	a chemical or natural substance added to soil to increase its ability to support plants

Continued

Table 4.8. (*continued*)

Key Vocabulary	Definition
gravity	the force that pulls a body toward the center of Earth or another physical body with mass
guard	to protect something valuable or precious
NOAA	the National Oceanic and Atmospheric Administration, a scientific government agency that studies the hydrosphere (atmosphere, oceans, and waterways) to inform the public about weather, fisheries management, safe navigation, and much more
perspective	a certain attitude toward or a way of thinking about something; a point of view
pesticide	a substance used to kill insects or other organisms that are harmful to plants or animals
phosphate	a family of chemicals used in fertilizer
photosynthesis	the process by which plants use sunlight to make food from carbon dioxide (CO_2) and water
public service advertising	advertising activities that are meant to educate the public about an issue rather than to sell products or services
ration	to allow people to use only a certain amount of a resource in order to conserve the resource
runoff	water that drains from land surfaces into ditches and streams and eventually flows into bodies of water such as rivers, lakes, and oceans
weather	the current outside conditions, whether hot or cold, wet or dry, and clear or cloudy
wick	a cord or strip of porous material through which a liquid is drawn by capillary action; also, to absorb or draw off (a liquid) by capillary action
xylem	plant tissue that moves water from the plant's roots through its stem and to its leaves

TEACHER BACKGROUND INFORMATION
Collecting Precipitation Data

This lesson uses rainfall statistics for your location. Detailed precipitation data are available on the National Oceanic and Atmospheric Administration (NOAA) National Weather Service website at *http://w2.weather.gov/climate*. The procedure for accessing rainfall statistics is provided on the Using the NOAA Website handout (p. 182).

Irrigation Techniques

Students investigate irrigation techniques in this lesson. The following is a summary of irrigation techniques used in agriculture:

- *Sprinkler.* Water is applied through pipes operated under pressure to form a uniform spray pattern.

- *Spray Irrigation.* Water is pushed out through a long tube that is affixed to a water source (perhaps a well) and is dispersed using a spray gun–like device.

- *Overhead Irrigation.* Water is piped to one or more central locations in the field and distributed by high-pressure sprayers or low-pressure sprinklers.

- *Center-Pivot Irrigation.* Pipes with sprinklers are supported above the crop by metal frames called towers located at fixed intervals. The pipe moves around a central pivot point on wheels, supplying water to the crops in a circular pattern at a uniform rate. The amount of water the crop receives is determined by the rate at which the pipes move.

- *Rotation Irrigation.* A system in which irrigators are provided an allotted amount of water at set intervals.

- *Drip Irrigation.* Water is applied directly into the root zone of the plants using applicators that operate under low pressure. The applicators are positioned either on the surface or below the ground.

- *Flood Irrigation.* The entire surface of the soil is covered by pond water.

- *Furrow (Ditch) Irrigation.* Water is applied in rows (furrows) that are deep enough to contain the specified amount of water causing the surface to be partially flooded.

- *Surface Irrigation.* The soil is used as the conduit because rainfall normally is sufficient for most of the watering needs.

- *Gravity Irrigation.* Water is not pumped through the system, but rather flows and is distributed by gravity.

- *Subirrigation.* By either raising the water table near the roots of the plant or burying porous pipes, water is supplied to plants underground and discharged directly into the root zone.

- *Terracing.* Large steps are cut into hillsides and supported by stone or concrete walls. The level parts are used as small fields, and as water flows down the hillside, it is channeled to each plot.

The following online resources provide additional information about irrigation techniques:

- "Irrigation," National Geographic: *www.nationalgeographic.org/encyclopedia/ irrigation*

- "Strategies for Efficient Irrigation Water Use," Oregon State University Extension Archive: *http://ir.library.oregonstate.edu/xmlui/bitstream/handle/1957/37465/em8783.pdf*

- "Irrigation System," Kids.Net.Au Encyclopedia: *http://encyclopedia.kids.net.au/page/ ir/Irrigation_system*

- "Irrigation Techniques," USGS Water Science School: *http://water.usgs.gov/edu/ irmethods.html*

- "Some Irrigation Methods," USGS Water Science School: *http://water.usgs.gov/edu/ irquicklook.html*

Agriculture and the Environment

This lesson aims to teach students about the undesirable consequences of pesticides and fertilizers without vilifying farmers. Agricultural practices such as irrigation and the application of pesticides and fertilizers help ensure that our nation has a consistent and abundant food supply; however, these practices can also result in pollution of nearby surface water and groundwater. The major sources of water pollution from agriculture are the addition of sediment and nutrients from fertilizers to waterways. Sediment runoff can cause turbidity (cloudy water), which can harm aquatic organisms; clog streams, drainage ditches, and water intake pipes; and make waterways less appealing for recreational uses. Fertilizer runoff adds phosphates, potassium, and nitrogen to waterways. These chemicals can promote the growth of algae and other plants in waterways, which may upset the balance of aquatic ecosystems.

Farmers employ several methods to reduce runoff from fields. These include leaving fields covered with residue from crops or allowing other plants to grow to reduce erosion when a field is not being actively used for growing crops, planting trees and shrubs at the margins of fields to prevent erosion and runoff, and strategically applying fertilizers and pesticides to crops to reduce the overall amount of chemicals used.

The following online resources provide an overview of the effects of agriculture on the hydrosphere and measures to protect the hydrosphere:

- "Battling the Bloom: Lake Erie": *www.youtube.com/watch?v=gMwQaHtK904&feature=youtu.be*

- Chesapeake Bay Program: *www.chesapeakebay.net/issues/issue/agriculture*

- "Toxic Algae Bloom Threatens Florida Waters": *https://www.youtube.com/watch?v=K_sLmKkC6yk*

COMMON MISCONCEPTIONS

Students will have various types of prior knowledge about the concepts introduced in this lesson. Table 4.9 outlines some common misconceptions students may have concerning these concepts. Because of the breadth of students' experiences, it is not possible to anticipate every misconception that students may bring as they approach this lesson. Incorrect or inaccurate prior understanding of concepts can influence student learning in the future, however, so it is important to be alert to misconceptions such as those presented in the table.

Table 4.9. Common Misconceptions About the Concepts in Lesson 3

Topic	Student Misconception	Explanation
Agriculture	Fertilizers help plants grow, so they are good for the environment.	Fertilizers applied to the soil can run off into groundwater and surface water and cause contamination by adding chemicals to water sources.
	All soil erosion is caused by human activities such as agriculture.	Soil erosion is caused primarily by wind, water, and gravity. Although human activities can increase or decrease the amount of soil erosion that occurs, erosion is a natural process.
Mathematics	A graph is a picture of a phenomenon. (For example, students may interpret a line graph with a peaked shape as a mountain and connect this perception with the phenomenon under investigation.)	Graphs show information mathematically, using data someone has collected. The shape of the graph may tell us something about the phenomenon, but it is not a picture of the phenomenon. It is important to read the axis labels to understand what the graph depicts.

PREPARATION FOR LESSON 3

Review the Teacher Background Information (p. 161), assemble the materials for the lesson, and preview the videos recommended in the Learning Components section below.

You may wish to invite the school custodian or building manager to visit the class to review students' maps to verify their accuracy and provide feedback. Display completed student maps around the classroom in a location where it will be easy for the custodian or building manager to review them.

If you choose to take the class on a field trip as part of the social studies connection, make arrangements with a local farm or greenhouse that uses practices such as sprinkling irrigation and water collection techniques and can demonstrate irrigation techniques and discuss waterway protection with students. Visit the site beforehand to check for potential hazards to students, such as trash, broken glass, rusted farm equipment, or poisonous plants. Alternatively, you might invite a guest speaker to talk to the class about irrigation and waterway protection measures.

In science, students will observe capillary action using ribs of celery. The ends of the celery ribs must be cut off to expose the xylem for this activity. You should prepare the celery in advance. Lay the individual ribs of celery on a cutting board, and cut each rib about 5 or 6 inches below where the rib and leaves meet. This will expose the xylem at the end of the cut rib. Cutting quickly and with a sharp knife will ensure that the openings of the tubes won't be compressed.

In ELA, students will be investigating the role of the written word in promoting opinions about public issues. They will create a public service advertising (PSA) campaign to encourage the community to protect the local watershed. Compile examples of how writing is used in PSA campaigns to persuade people to adopt a certain opinion about an issue. These may include brochures promoting healthy eating habits, books about social issues, pictures of billboards, websites about conservation issues, and examples of PSAs that are available online (these are based on written scripts). The following web pages contain information about and examples of PSA campaigns that you may wish to share with students:

- *www.adcouncil.org/Our-Campaigns/The-Classics*

- *www.thebalancecareers.com/what-exactly-is-public-service-advertising-38455*

LEARNING COMPONENTS
Introductory Activity/Engagement

Connection to the Challenge: Begin each day of this lesson by directing students' attention to the driving question for the module: How can we use what we know about rainfall to design a system to provide water to a garden? Hold a brief student discussion of student prior knowledge on this topic, creating a class list of key ideas on chart paper, or you may wish to have students create a notebook entry with this information.

Driving Question for Lesson 3: What principles and techniques are needed to design a water collection system that handles and stores irrigation water efficiently?

Mathematics Class and Science and Social Studies Connections: First, hold a class discussion about the term *guard*, asking students to define the word. Students will likely reference the term's meanings (e.g., to prevent a prisoner from escaping, to protect an area from an enemy, a protective part of a piece of equipment, to safeguard something valuable or precious). Focus students' attention on the concept of safeguarding something valuable or precious, and ask students what valuable resource they are learning about in the module (water). Next, show the Google slideshow "Guarding the Hydrosphere," at *http://tinyurl.com/GuardHydrosphere*, stopping to discuss the questions and view the video.

Tell students that in their Rainwater Roundup Challenge, the fictional town of La Vieja is experiencing a water shortage, and therefore water must be rationed. As a class, discuss the concept of shortages, asking students if they have ever experienced a time when something they needed or wanted was in short supply. Hold a class discussion about the role of scientists, communities, and citizens in guarding water resources. Emphasize to students that although only some parts of the United States are suffering from water shortages, water is a valuable resource everywhere because it is necessary for all life.

STEM Research Notebook Prompt

Have students respond to the following prompt: *How can you guard our community's water resources?*

Remind students about the objective of the module (providing water for the garden at Sunny Acres). As a class, discuss the role of water in the Sunny Acres story, asking question such as these:

- How important is water to maintaining a garden like the one Grandpa Henry tends?

- Who made the decision that La Vieja water supplies cannot be used for lawns or gardens?

- Who benefits from this decision?

- Why is it important for the people who live in the La Vieja community to heed this decision to monitor how water is used?

ELA Connection: Hold a class discussion about how Rachel Carson guarded a part of the natural environment by writing about it, asking students if they can think of other examples of how people have used writing to protect the environment. Encourage students to think about brochures, billboards, and letters to the editor in the newspaper. Even televised PSAs can be examples of using writing to persuade people to agree with a certain opinion, because they are based on written scripts. Show the class examples of texts and advertisements that are used to disseminate an opinion to the public (see Preparation for Lesson 3 on p. 164). Hold a class discussion about how students have experienced this sort of persuasive text, creating a class list of students' ideas and experiences.

Activity/Exploration

Mathematics Class: An option for this lesson is a visit by a school custodian or building manager to review teams' schoolyard maps and give students an insider's view of the different systems used in the schoolyard to handle rainwater. The custodian or building manager may be able to give information about elevations of different parts of the schoolyard, for example. This will help students understand the flow of rainwater from the school property. He or she may also provide insight about the building's structure that students may not have observed when they gathered their data.

What's the Volume?

Reread the Sunny Acres design challenge scenario originally presented to the class in Lesson 1 (Rainwater Roundup handout, p. 84). Ask students what kind of information they need to design a rainwater-recycling system to help Grandpa Henry and Mamito Anna save the garden. Create a class list of student ideas.

Extend students' thinking about rainfall and the rainwater-recycling system by discussing the following questions as a class:

- Where does the precipitation that falls to Earth go?

- Can more rain fall in one area than in another area if they are in the same general vicinity?

- If water falls to Earth's surface and enters the ground (groundwater and other bodies of water), what happens to the water that falls on structures and hard surfaces such as the playground or parking lot?

- How could knowing the size of the playground or the school building help us design a water collection system? (Is the total volume of rainfall in the layer of water that collects on this surface important? Why or why not?)

Ask students if they think that the time in which an amount of rain falls matters for directing rain away from a building and for collecting rainwater. Introduce the idea that rainfall intensity is classified by the amount of rain that falls during an hour:

- Light rain = less than 0.1 inches per hour

- Moderate rain = between 0.1 and 0.4 inches per hour

- Heavy rain = between 0.4 and 2.0 inches per hour

- Violent rain = greater than 2.0 inches per hour

Next, sketch a diagram of the school playground on the board. Ask students to predict the total water volume based on 0.25 inches (¼ inch) of rainfall. Remind students that they calculated the volume of rain for the school building in the last lesson, encouraging them to consider how the area of the playground compares with the area of the school building.

Distribute the What's the Volume? handout (p. 178) to each team, and have teams calculate the total volume of rainwater that falls on the playground during a light rain lasting most of the morning. For this activity, the TV news weather segment reports the day's rainfall at 0.25 inches.

After teams have completed their calculations, have them share their findings with the class. Compare these findings with the volume of rain for the school building that teams calculated in Lesson 2. Remind students that in the Rainwater Roundup Challenge scenario, their school playground is about the same size as the parking lot at Sunny Acres. Have students suggest possible ways for collecting the water (e.g., an underground tank fed by a drain at a low spot in the playground, a tank placed at the end of a drain from the playground). Have teams share their ideas with the class, and tell students that their teams will design rectangular or cylindrical tanks that can hold the volume of water that they calculated for the playground.

STEM Research Notebook Prompt

Students should respond to the following prompt: *How did your calculation compare with the prediction? Write a set of instructions for calculating rainfall volume on a large surface that is not rectangular.*

How Big Is Big Enough?

Tell students that they will begin making decisions about their water collection system's design. Hold a discussion to guide students to consider the shape and size of the collection system, asking the following questions:

- What kinds of structures have you seen in the community that hold water?

- How would you describe the structures you've seen (round or square, large or small, on the ground or in the air)?

- What might be the way people get the water from these structures (e.g., using a faucet, a hose)?

- As you consider the size and design for your own water collection structure, you will need to know how much water it should store. How could you find out how much water it would take to water the garden at Sunny Acres?

Read aloud to the class the Letter From Mamito Anna (p. 179). This message gives new information about potential locations for the rainwater-recycling tank (in her letter, Mamito Anna suggests two alternatives for the location of a storage tank). Have teams discuss rainwater-recycling options, considering how this new information might change the ideas they discussed earlier about where and how to collect water. Next, ask for students' opinions about whether a storage tank is needed. Review the total rainfall volumes the design teams calculated. Have students venture opinions about the following questions related to the quantity of water needed:

- Which of these containers might hold the calculated amount of water: a bathtub, a wading pool, a kiddie pool, a park swimming pool, a water tower?

- How much water is needed to water the garden at Sunny Acres?

- Why would you want to use less water?

- How could you minimize the amount of water needed for the garden?

Review the calculations for determining the volumes of rectangular solids ($V = l \times w \times h$) and cylinders ($V = \pi \times r^2 \times h$), working through examples as a class. Distribute the How Big Is Big Enough? handouts to each team. Have each design team determine the feasibility of storing the calculated volume of water in either of the locations Mamito Anna proposed. After selecting the location for the tank, teams will each design a rectangular or cylindrical tank that can hold the calculated amount of water.

When teams have finished calculating the dimensions, have each team glue a construction paper footprint onto its Schoolyard Map to represent the tank's location. The construction paper footprint does not need to be scaled to the map, although its size should approximately fit the location the team chose for the tank on the map. Then, the team should write the tank's dimensions (height and total capacity in liters) on a notecard and glue it onto the construction paper footprint.

STEM Research Notebook Prompt

Students should respond to the following prompt: *Can the tank fit the spots Mamito Anna suggested? Why or why not? What problems will need to be solved to use either of these locations?*

Science Connection: Introduce the topic of capillary action by holding a class discussion about the behavior of water. Remind students that they have been observing the way water behaves in their biodomes and in their schoolyard. Ask students to share their

observations. Next, ask students if water always flows downhill, asking them to justify their reasoning (e.g., water always flows downhill because of gravity).

Way to Flow: Capillary Action

Begin the activity by asking students if they have ever wondered how water gets from the roots of a tree into its leaves. Define the term *capillary action* and explain to students that capillary action causes water to flow *uphill* from a plant's roots to its leaves. Explain to students that they see capillary action at work when they use a paper towel to soak up spilled water. Pass out the Way to Flow: Capillary Action handout to each student and review the instructions. Students will make periodic observations and record information on a data table on the handout, as well as answer the following questions in their STEM Research Notebooks.

STEM Research Notebook Prompt

Have students respond to the following prompts:

- *Before making your first observation, write down your predictions: What do you think will happen? When will the greatest amount of change take place?*

- *After your observations, answer these questions: What did you observe? How fast did the water climb the celery? What happened to the colored water as time went by? Explain.*

- *Describe how farmers can use their understanding of capillary action and the attraction of water molecules to each other and other substances to improve the health of their crops.*

ELA Connection: Tell students that they are going to use persuasive writing to communicate messages to the community about caring for the local watershed. Remind students of Rachel Carson's work, and ask students for their ideas about what she did first, before she wrote her book *Silent Spring*. Prompt students to understand that she first identified a problem (birds dying) and then conducted research. Remind students that they have been conducting research about the watershed in mathematics, science, and social studies classes and that their STEM Research Notebooks will be an important source of information as they create their community messages.

Introduce the idea of PSAs as a kind of advertising that is meant to educate the public about an issue rather than to sell them something. Review one or two of the examples of PSAs you shared with students in the Introductory Activity/Engagement ELA activity (see p. 165), and ask students what they notice about the kind of language used. Introduce the idea that persuasive writing is a kind of writing in which the writer aims to influence the opinion of the reader by stating an opinion and providing reasons why the

reader should adopt this opinion. Tell students that they will create a PSA campaign to share their opinions about watershed protection to influence the public.

First, however, tell students that they need to define the problem they will address. Remind students of the slogan they created for watershed protection in Lesson 1. Ask students if this slogan contains a statement of the problem (slogans are typically statements to compel action rather than to define a problem). Have students work as teams to create a one-sentence problem statement. Have teams share their problem statements, and then have other teams give feedback on the statements. As a class, combine elements of teams' problem statements to create one problem statement for the class. Next, introduce the idea that public service messages usually have an action element associated with them (i.e., what the advertiser wants people to do). Have teams work together to create action statements by completing this sentence: "To address this problem, people should …" The slogan developed in Lesson 1 may be aligned with an action. However, it is likely that the slogan does not contain specific action items. Encourage teams to provide specific details in their action statements rather than restating the slogan. Have teams share their action statements, and then as a class, develop a single action statement.

Ask students to compare their problem and action statements with the slogan from Lesson 1 and decide as a class if they believe that the slogan is still the best one to use. Remind students that the slogan should be short and memorable and should address the issue without providing too much detail. Tell students that the goal of the slogan is to catch peoples' attention so that you can provide them with more specific information. If students believe the slogan should be modified, work as a class to incorporate students' ideas to modify the slogan or create a new slogan.

Tell students that the next step is to provide reasons why people should agree with the problem statement and take the actions suggested. Introduce the idea that the word *OREO* can help them remember the steps in writing an opinion paper. Pass out the Writing the OREO Way handout (p. 187), and review the components of opinion writing outlined on the handout. The OREO writing steps are as follows:

- O = Write an introduction clearly stating your *opinion* (problem statement and action statement).

- R = Give two or three *reasons* why you hold this opinion.

- E = Provide two or three *examples* that support these reasons, providing details.

- O = Restate your *opinion* in your conclusion.

Have students complete the Writing the OREO way handout individually, reminding them to use their learning in mathematics, science, and social studies and their STEM Research Notebooks as sources of their reasons and examples.

Social Studies Connection: Introduce the idea to students that water is a public good, something provided to people for their well-being by the government. Ask students for their ideas about who makes decisions about how water is used in their community, guiding students to understand that this is usually a function of government officials. Next, ask students to share their ideas about what kinds of decisions these people have to make (e.g., how much water will be provided for crops, how much for drinking water, and how much for businesses to use in manufacturing). Hold a class discussion, asking students to respond to the following:

- What are some ways that water is used in a community? (factories, swimming pools, greenhouse watering systems, drinking, bathing, cooking, etc.)

- How do community decision makers decide how water is used?

- In La Vieja, the city government has decided that water can no longer be used for watering gardens and lawns. Why has it made this rule? Do you think it is fair?

Show a video that illustrates the importance of water and irrigation strategies to agriculture, such as "Wise Water Use—Explaining Irrigation" at *www.farmingismagic.co.uk/films/wise-water-use*.

Irrigation Research

Have students choose 10 irrigation strategies from the following list to research and evaluate for the effectiveness and wise use of water resources:

- Watering with a garden hose
- Center-pivot irrigation
- Rotation irrigation
- Sprinkler
- Drip irrigation
- Furrow irrigation

- Flood irrigation
- Surface irrigation
- Ditch or subirrigation
- Gravity irrigation
- Terracing
- Overhead irrigation

Students should enter the following information about each strategy, along with a labeled sketch, in their STEM Research Notebooks:

- Name of the irrigation strategy
- How the irrigation strategy works to deliver water
- Advantages of this irrigation strategy
- Disadvantages of this irrigation strategy

- Rating for how they would recommend this irrigation strategy to farmers (on a scale of 1–10, with 1 being the strategy they would most recommend that a farmer use)

Some suggested online resources on irrigation techniques are listed in the Teacher Background Information section (see p. 162). After researching and rating the irrigation systems, students should write a summary paragraph explaining their reasoning for their ratings. Have students share their results with the class and defend their positions.

Farm or Greenhouse Field Trip or Guest Speaker (*Optional*)

Students have learned about the importance of water and using effective irrigation strategies as well as being wise stewards of the environment. A visit to a farm or greenhouse will help them make connections between their learning and the real world. An alternative option is to have a farmer or greenhouse worker visit the classroom to speak to students. Students should be aware that most farmers and horticulturists are concerned about the environment and seek ways to protect the groundwater from harmful fertilizers and pesticides. Before the field trip or guest visit, have teams work together to prepare two or three questions they would like to ask the farmer or greenhouse worker.

Explanation

Mathematics Class: In this activity, student teams analyze annual rainfall data for their local area and use these data to determine the rainwater supply for the recycling system.

Understanding Weather Data

Introduce the activity by discussing weather and climate with the class, using the following prompts and questions:

- How does your family get information about the weather?

- Why is it important to know about the weather?

- Describe the weather this week, day by day.

- Why is information about weather in the past important?

- From our story about Mamito Anna and Grandpa Henry, what do we know about the climate in La Vieja, California?

Then, demonstrate how to use the NOAA website and how the data are displayed, following the instructions on the Using the NOAA Website handout. Have students work in teams to record the historical precipitation data they find on an electronic spreadsheet or in a table they prepare in their STEM Research Notebooks. Following is an example of what this table should look like, with a title and these headings over each column. Students should create a row for each month.

Annual Rainfall Statistics Monthly Data Table

Month and Year	Month Total (inches)	Average (inches)	Departure From Average (inches)	Previous Year (inches)
December 2017	2.96	3.10	−0.14	1.95
January 2018	2.40	2.96	−0.56	3.12

Using these data, have students create a bar or line graph in their STEM Research Notebooks showing relative monthly precipitation amounts across the entire year. This will help students identify the rainiest months of the year.

Science Connection: Review the concept of capillary action, asking students to provide some examples of where they think they have experienced capillary action in everyday activities (e.g., paper towel soaking up water; watering plants at home; crying, because tears are an example of capillary action).

The Great Escape: Capillary Action

Introduce the idea that capillary action occurs because the particles that make up water are "sticky." Introduce the terms *cohesion* (water molecules are attracted to each other) and *adhesion* (water molecules are attracted to other substances). Tell students that not only does water tend to stick to itself, like in a water drop, but it also sticks to other substances such as a cup, cloth, soil, and other materials. Distribute the Great Escape: Capillary Action handout to each student and review the instructions. Show students how to create a wick by twisting a whole paper towel until it forms something that looks like a piece of rope.

STEM Research Notebook Prompt

Have students respond to the following prompts:

- *Before making your first observation, write your predictions in your STEM Research Notebook. What do you think will happen to the water in the cup with water? What will happen in the cup without water? Explain your reasoning about why you think this will happen. Use key vocabulary terms such as* capillary action, attraction, *and* force.

- *After your observations, draw a picture of what happened, using arrows to show direction and labels to identify forces. Explain your drawings and support your ideas with evidence.*

- *Then, describe how this activity demonstrates that chemicals in the water might have an impact on other regions that do not directly receive the chemicals that farmers apply. Consider possible alternatives that could be implemented.*

Have students share with the class their ideas about how chemicals that farmers and gardeners use on the soil enter plants. Prompt students to consider that there may be chemicals that are not absorbed into the plants that may move into watersheds along with excess water in the soil.

ELA Connection: Have students continue working on their Writing the OREO Way handouts as the basis for their PSA campaign. Once students have completed their handouts, have teams work together to compare team members' reasons and explanations and identify which ones they think are the best, choosing two or three reasons and two or three explanations. Based on these choices, have each team complete another Writing the OREO Way handout that incorporates the teams' best ideas.

Social Studies Connection: If students went on a field trip or a visitor spoke to the class about farming or greenhouse work, have students complete a STEM Research Notebook entry.

STEM Research Notebook Prompt

If students took a field trip, have them respond to the following prompts:

- *Write about any observations at the farm or greenhouse that were surprising to you.*

- *List methods used to irrigate and protect crops from pests.*

- *What watering collection system was used?*

- *Draw and label the watering collection devices you saw.*

- *What other equipment did you see? What was it used for?*

- *Identify other areas in which the farmer or greenhouse worker used science principles or research to improve his or her farming or gardening techniques.*

If a guest speaker visited the class, have students respond to the following prompts:

- *Describe the kind of work the guest speaker does.*

- *What did you learn about irrigation and watering from the speaker?*

- *What was the most interesting or surprising thing you learned from the speaker?*

Elaboration/Application of Knowledge

Mathematics Class and Science Connection: Ask students if they think that their rain gauge data can be compared with the NOAA weather data they collected. Point out to students that the data they collected from NOAA is for a month's rainfall, but they have not yet collected rainfall data from their rain gauges for an entire month. Introduce the idea that a way to make sure that their data are valid is to compare them with other measurements. Working in teams, students should compare their rain gauge data with the data from at least one other team. Hold a class discussion about whether teams' data were similar, creating a class chart of each team's rain gauge measurements. Next, have students compare their rain gauge data with day-by-day NOAA statistics following the procedure described on the Using the NOAA Website handout.

STEM Research Notebook Prompt

Have each student respond to the following prompt: *How did your team's rain gauge data compare with those of other teams? What are some explanations for differences between your team's rain gauge measurements and data from the other teams? How did your rain gauge measurements compare with data from the weather service? Give possible explanations for differences between your team's rain gauge measurements and the NOAA data.*

ELA Connection: Tell students that they will work as a class to spread the message about watershed protection to the community in a PSA campaign, but that each team will use a different method for spreading the message. Remind students of the different ways that public service messages were communicated, using the examples you provided in the Introductory Activity/Engagement section (e.g., brochures, websites, television advertisements, billboards or posters, radio announcements). As a class, decide what forms of communication you wish to use in the class PSA campaign. You may wish to provide a list of media from which students can choose, such as television advertisement, radio advertisement, newspaper advertisement, brochure, billboard, or website (content on the class website or a mock-up of a website). Ensure that you have access to technology and materials for each type of media the class may choose. Each team will be responsible for one method of communication. Work as a class to decide what teams will be responsible for each type of communication. For example, one team might create content for the class website or create a blog, while other teams create a "billboard" (poster), a brochure, and a television advertisement.

Once each team has been assigned a form of communication, have teams share their Writing the OREO Way team handouts with the class. As a class, compare and contrast the reasons and explanations teams provided to support their opinions. Tell students that although their PSA campaign will use different ways of spreading the message about watershed conservation, it is important to have a consistent message throughout

the different forms of communication. Work together as a class to create a single message using the Writing the OREO Way format. Record the class's decisions about the content on chart paper, and post it where it will be visible to students for the remainder of the module.

Social Studies Connection (*Optional*): After the field trip or visit from a guest speaker, have the class prepare a thank-you poster, which might include drawings of what they saw during the trip and messages of appreciation.

If your class went on a field trip, work together to identify how each of Earth's spheres was represented at the farm or greenhouse and how these spheres interacted. Create a Venn diagram on chart paper based on student responses, showing the spheres and their interaction.

Evaluation/Assessment

Students may be assessed on the following performance tasks and other measures listed.

Performance Tasks

- What's the Volume?
- How Big Is Big Enough?
- Understanding Weather Data
- Irrigation Research Rubric
- Way to Flow: Capillary Action
- The Great Escape: Capillary Action
- Writing the OREO Way Rubric

Other Measures

- STEM Research Notebook entries
- Engagement in class activities and discussions
- Involvement in group work and discussions

INTERNET RESOURCES

NOAA's National Weather Service climate data
- *http://w2.weather.gov/climate*

Irrigation techniques
- *http://education.nationalgeographic.org/encyclopedia/irrigation*

- *http://ir.library.oregonstate.edu/xmlui/bitstream/handle/1957/37465/em8783.pdf*

- *http://encyclopedia.kids.net.au/page/ir/Irrigation_system*

- *http://water.usgs.gov/edu/irmethods.html*

- *http://water.usgs.gov/edu/irquicklook.html*

Effects of agriculture on the hydrosphere and protection measures
- *www.youtube.com/watch?v=gMwQaHtK904&feature=youtu.be*

- *www.chesapeakebay.net/issues/issue/agriculture*

- *https://www.youtube.com/watch?v=K_sLmKkC6yk*

PSA campaigns
- *www.adcouncil.org/Our-Campaigns/The-Classics*

- *www.thebalancecareers.com/what-exactly-is-public-service-advertising-38455*

"Guarding the Hydrosphere" slideshow
- *http://tinyurl.com/GuardHydrosphere*

"Wise Water Use—Explaining Irrigation"
- *www.farmingismagic.co.uk/films/wise-water-use*

IMAGE CREDITS FOR LESSON 3

Water tank (p. 180)
www.flickr.com/photos/35090117@N05/4780424856
Owner: VelaCreations; License: Attribution 2.0 Generic (CC BY 2.0)

NOAA screenshots (pp. 182–184)
http://w2.weather.gov/climate
Owner: National Oceanic and Atmospheric Administration; Fair use

Team Name: _____   Date: _____

WHAT'S THE VOLUME

Your team will design a tank big enough to hold the rain that falls on the playground in one morning.

HOW MUCH WATER FELL ON THE PLAYGROUND?

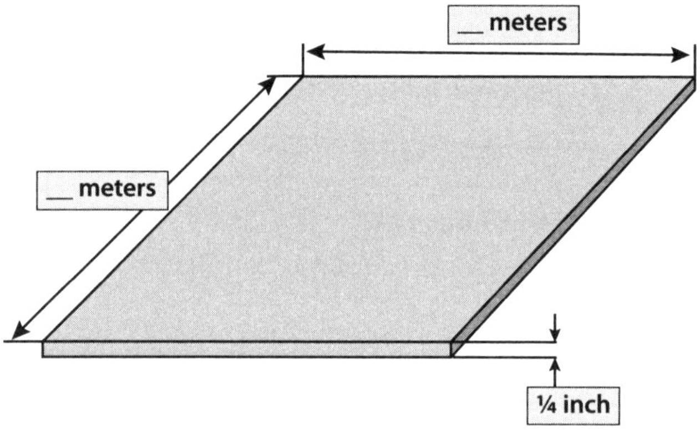

Last week, the class surveyed the schoolyard. From the data gathered, your team created a map of the schoolyard showing the school building, playground, and parking lots. On a rainy day, the same amount of rain falls everywhere on the schoolyard, unless an area is covered by a roof or sheltered by trees.

__ meters
__ meters
¼ inch

In the United States, rainfall is measured in inches. A light rain falling for a few hours could drop between 0.25 and 0.5 inches of rain. For this activity, calculate the volume in liters of rainwater that falls on the playground when there is 0.25 inches of rain.

Predict the playground's water volume at 0.25 inches of rain:
Calculate the area of the playground in square meters:
Convert 0.25 inches to meters:
Calculate the volume of a 0.25-inch layer of water in cubic meters:

STUDENT HANDOUT

LETTER FROM MAMITO ANNA

From the Desk of Mamito Anna

Sunny Acres | La Vieja, CA

March 1, 2019

Dear Jorge and Angie and friends,

Thanks for the great news that your class is working on a plan to save our garden. I told Henry about your ideas, and he was all smiles! I think it actually helped him do better on his painting lessons!

When we mentioned your idea to Kate, the fix-it person at Sunny Acres, she said that the big washing machine is being removed from the utility room. Could we put the rainwater tank there? Henry also suggested that we might be able to move the garden shed and use that spot for the tank. Honestly, I don't know why he keeps that old shed! There's nothing in it but a bunch of old flowerpots. Henry said to tell you it's 8 feet by 10 feet.

Love,

Mamito Anna

Team Name: _____   Date: _____

HOW BIG IS BIG ENOUGH?

Your team will design a tank big enough to hold the rain that falls on the playground in one morning.

STORAGE TANK: HOW BIG IS BIG ENOUGH?

Earlier, your class calculated the amount of water on the school building and on the playground with 0.25 (¼) inch of rainfall. Were you surprised at the amount of water that fell? In this activity, you will design a tank that will hold the amount of water you calculated for the playground.

Read Mamito Anna's note, which provides some ideas she had for where to place the tank. Remember, the playground at your school is just about the same size as the parking lot at Sunny Acres. Using mathematics, determine if either of Mamito Anna's ideas could work.

How big a tank would fit where the washing machine was sitting?
• A typical large-capacity washing machine is about 27 inches wide, 40 inches tall, and 33 inches deep. • Calculate the volume of a tank that size in liters. • Would it hold the rainwater?
Show the dimensions and your calculations here:
How big a tank would fit where the garden shed is sitting now?
• Calculate the maximum area of the shed location in square meters. • How tall would the tank need to be to hold the water? • What other dimensions could you use for a tank where the area of the base is smaller?
Show the area and your calculations here:

Team Name: _____   Date: _____

HOW BIG IS BIG ENOUGH?

Design a tank to hold 0.25 inches of rainfall covering the surface of your playground. The tank can be a cylinder or a rectangular solid.

- Make a sketch of the tank in the space below.
  - *Rectangular Tank:* show front and side views with dimensions for length, width, and height
  - *Cylindrical Tank:* show front and top views with dimensions for height and diameter
- Be sure to note the material to be used for the tank and its total capacity in liters.

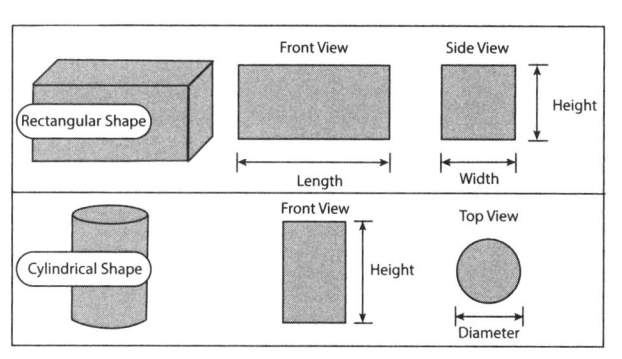

Tank Notes:

USING THE NOAA WEBSITE

ANNUAL RAINFALL STATISTICS FOR PRECIPITATION ANALYSIS

1. Visit the National Weather Service Climate Services web page at *http://w2.weather.gov/climate*.

2. Click on the region of the state where your school is located.

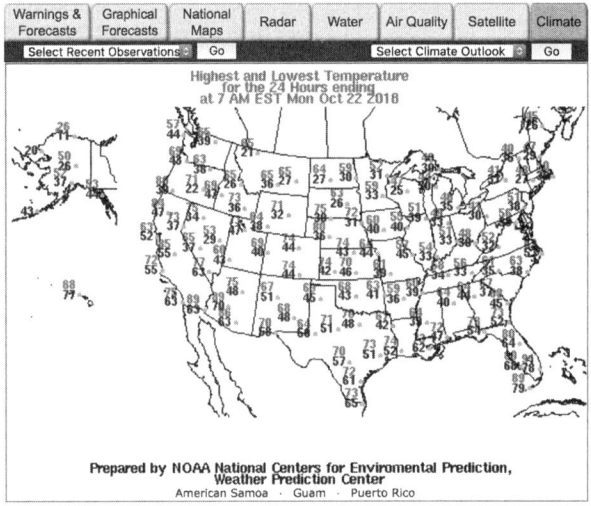

3. On the "Observed Weather Reports" page that comes up for the region:

    -   Select "Monthly Weather Summary (CLM)."

    -   Select the nearest location. (If you don't see your city or town listed here, clicking on the "Climate Locations" tab will show you all the local reporting stations available to choose from. You may have to choose another town that's as close to yours as possible.)

    -   Choose "Archived Data," scroll down to highlight January of the previous year, and then click "Go."

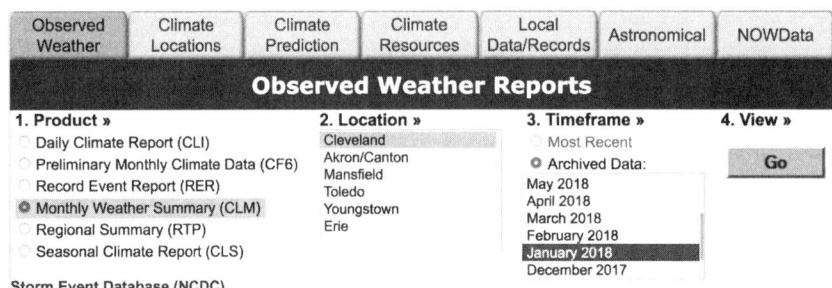

STUDENT HANDOUT, PAGE 2

USING THE NOAA WEBSITE

4. This will bring up that month's Climatological Report (CLM) as a simple web page of weather data. (See the graphic below for an example of the data shown on a CLM.)

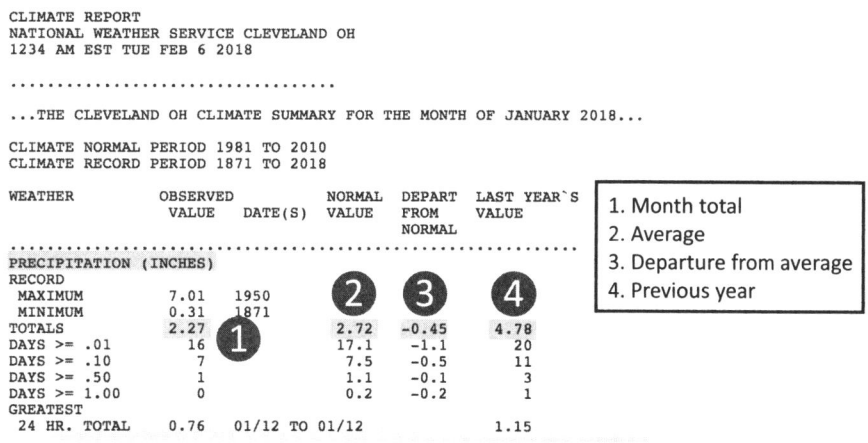

5. Follow the same procedure month by month to collect precipitation data for January through December of the previous year.

6. As you access each CLM, post the data for each month to a table on an electronic spreadsheet such as Google Sheets or Excel. Enter data as they are in the table, but set the cells to round to the nearest tenth of an inch.

PAST MONTH RAINFALL STATISTICS FOR RAIN GAUGE VALIDATION

1. Visit the National Weather Service Climate Services web page at *http://w2.weather.gov/climate.*

2. Click on the region of the state where your school is located.

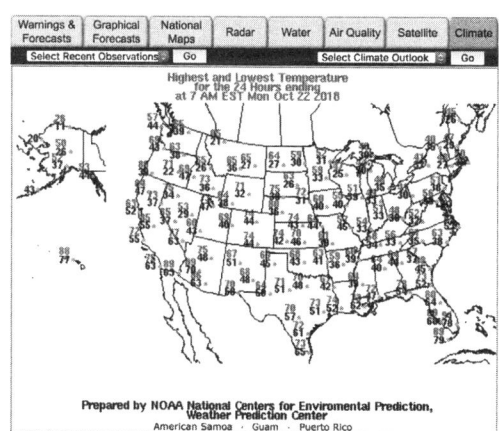

STUDENT HANDOUT, PAGE 3

USING THE NOAA WEBSITE

3. On the "Observed Weather Reports" page that comes up for the region:
   - Select "Preliminary Monthly Climate Data (CF6)."
   - Select the nearest location.
   - Click "Most Recent" and then "Go."

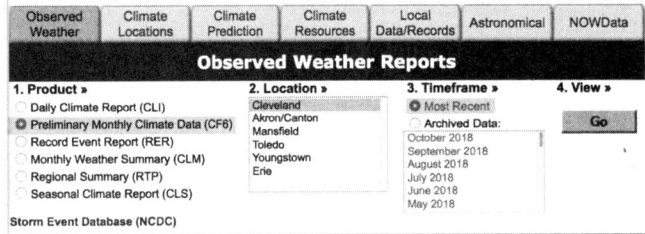

4. The current month's Preliminary Monthly Climate Report (CF6) comes up in a separate web page.

5. Collect precipitation data for each day to compare against your team's rain gauge data results.

6. If your team's rain gauge data extend back to the previous month, pull the previous month's report also (see "Choose previous month" on the graphic).

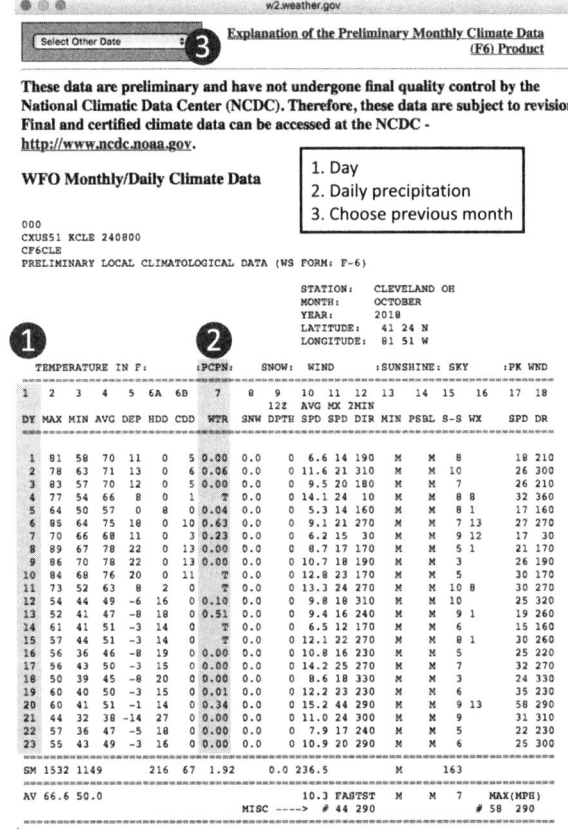

4

Name: _____   Team Name: _____

WAY TO FLOW: CAPILLARY ACTION

Water can move in mysterious ways. It is one of the few substances on Earth that can defy gravity. We are going to use ribs of celery to see this amazing phenomenon.

DEFYING GRAVITY WITH CAPILLARY ACTION

1. Place the celery on a paper towel and observe the cut ends of your celery to identify the xylem. Create a labeled sketch in your STEM Research Notebook with colored pencils or markers.

2. Fill the beaker or cup halfway with cold water.

3. Add 20 drops of food coloring to make the water a deep color.

4. Put the celery ribs inside the beaker or cup, with the cut ends in the water.

Make observations using the data table below. Measure the distance the colored water has traveled, and record this amount. During each observation, show on your sketch how far up the celery rib the food coloring has traveled.

Time	The celery looks like this ...		
30 min.		1 hr.	
2 hrs.		3 hrs.	
4 hrs.		6 hrs. or next day	

Name: _____   Team Name: _____

STUDENT HANDOUT

THE GREAT ESCAPE: CAPILLARY ACTION

Capillary action is not limited to plants; it occurs because water is "sticky." Water uses the forces of cohesion (water molecules prefer to stay close together) and adhesion (water molecules are attracted and stick to other substances). Not only does water tend to stick together in a drop, but it sticks to other materials as well.

1. Use a permanent marker and your ruler to draw 8 to 10 evenly spaced marks on each cup.

2. Fill one cup with water to the top mark, leaving the other empty.

3. Twist a paper towel until it forms something that looks like a piece of rope; this will be the wick.

4. Place one end of the wick into the cup that is filled with water, and place the other end into the empty cup.

5. Watch what happens over time, and record your observations on the illustrations below. At each interval, draw a picture of the wick that is spanning the cups and the approximate amount of water in each cup.

The water and wick look like this ...				
30 min.			1 hr.	
2 hrs.			3 hrs.	
4 hrs.			6 hrs. or next day	

Name: _____   Team Name: _____

STUDENT HANDOUT

WRITING THE OREO WAY

Opinion	Give your opinion (state the problem and the actions people should take).
Reason	State the reason for having this opinion.
Example	Give an example that supports your opinion.
Reason	State a second reason for having this opinion.
Example	Give a second example that supports your opinion.
Opinion	Restate your opinion.

Writing the OREO Way Rubric

Name: _____

Team Name: _____

Criteria	Exceeds Standard (4 points)	Meets Standard (3 points)	Approaching Standard (2 points)	Needs Improvement (1 point)	Score
SUPPORT FOR OPINION	Opinion is clearly stated and well-supported by two convincing examples.	Opinion is stated and 2 examples are provided.	Opinion is stated and 1 example is provided.	Opinion is not stated or is unclear and/or no examples are provided.	
ORGANIZATION	Work is extremely neat and organized.	Work is neat and organized.	Work is somewhat neat but lacks organization.	Work is not neat or organized.	
MECHANICS AND SPELLING	No spelling or grammar errors.	Few spelling and grammar errors.	Some spelling and grammar errors.	Many spelling and grammar errors.	

TOTAL SCORE: _____

COMMENTS:

Irrigation Research Rubric

Name: _____

Criteria	Exceeds Standard (4 points)	Meets Standard (3 points)	Approaches Standard (2 points)	Needs Improvement (1 point)	Score
CONTENT	Student recorded details about 10 irrigation systems, including clear and thorough descriptions of the systems and 2 or more advantages and disadvantages for each. There is evidence that the student used the findings to rank the systems.	Student recorded details about 10 irrigation systems, including clear descriptions of the systems and 1 advantage and disadvantage for each. There is evidence that the student used the findings to rank the systems.	Student recorded details about 7 or fewer irrigation systems; descriptions of the systems may not be clear and/or no advantages and disadvantages are listed. Student ranked the systems, but there is little evidence that these rankings are based on research findings.	Student recorded details about 5 or fewer irrigation systems; descriptions of the systems may not be clear and/or no advantages and disadvantages are listed. Systems may not be ranked, or rankings are unclear.	
SKETCH	Neat, labeled, and easy to understand sketches are provided for 10 irrigation systems.	Labeled sketches are provided for 10 irrigation systems.	Sketches are provided for some, but not all, irrigation systems described and may not be neat.	Few sketches are provided, and sketches lack labels and/or are not understandable.	
GRAMMAR AND SPELLING	There are no grammar mistakes and no spelling errors.	There are one or two grammar mistakes or spelling errors.	There are three to five grammar mistakes or spelling errors.	There are more than five grammar mistakes or spelling errors.	

TOTAL SCORE: _____

COMMENTS:

Lesson Plan 4: The Rainwater Roundup Challenge

This lesson begins with a panicked Angie sending the teacher an e-mail for the class. She just realized they have a big problem with the rainwater-recycling project … gravity! Students are challenged to help her create solutions for this problem. They create proposals for solving the problem and present them to Sunny Acres stakeholders in an oral presentation.

ESSENTIAL QUESTIONS

- How can we use data to calculate water needs?

- What makes a surface a good place to collect rainwater for recycling?

- Is it possible to make water flow uphill? If so, how?

- What evidence do we have to support our design decisions for the Rainwater Roundup Challenge?

ESTABLISHED GOALS AND OBJECTIVES

At the conclusion of this lesson, students will be able to do the following:

- Use a spreadsheet to create a way to conduct repetitive calculations

- Calculate the volume of rectangular prisms and cylinders of given dimensions

- Design a system to store a specified amount of water

- Use their understanding of irrigation techniques to propose an irrigation system for a garden

- Create a model of an irrigation system

- Synthesize their learning from the module to create a model of a rainwater capture system that includes rainwater storage and a system to deliver stored rainwater to a garden

- Create a slideshow presentation for their rainwater capture system and the process they used to create the system

- Synthesize their learning about watershed protection to create persuasive text to convey a message about watershed protection to the public in a media format

TIME REQUIRED

- 6 days (approximately 45 minutes each day; see Tables 3.9 and 3.10, pp. 45–46)

MATERIALS

Necessary Materials for Lesson 4

- STEM Research Notebooks

- Internet access

- Handouts and rubrics (attached at the end of this lesson)

Additional Materials for Mathematics Class and Science Connection (per team unless otherwise noted)

- Software or online program for spreadsheets such as Google Sheets or Microsoft Excel (per class)

- Google Slides or PowerPoint software

- 3 paper or foam cups (12 oz.)

- 30 coffee stirrers

- 30 flex straws

- 10 paper clips

- 40 craft sticks

- Hot glue gun

- Glue sticks

- Glue

- Duct tape

- 11 oz. nonhardening modeling clay

- 5 mm foam board (20 × 30 inch sheet)

- 2 shoe boxes (or similar-size cardboard boxes)

- 1 m aquarium flexible tubing (3/16 or 3/8 inch diameter)

- 10 #33 rubber bands

- 2 foam egg cartons (1 dozen size)

- Indirectly vented chemical splash goggles (1 per student)

- Piece of copy paper (1 per student)

- Angie's E-mail With Playground Sketch handout (1 per student, see p. 210)

- Cylinder Volume Spreadsheet Example handout (p. 211)

- Angie's Garden Sketch handout (p. 212)

- Garden Water Investigation handout (p. 213)

- Rainwater Roundup Challenge Checklist (1 per student; see p. 214)

- EDP Applied to the Rainwater Roundup handout (1 per student; see p. 215)

- Building an Argument handout (see p. 216)

Additional Materials for ELA and Social Studies Connection (materials needed depend on class decisions about media for their PSAs)

- Technology to video record student "television" PSAs

- Technology to audio record student "radio" PSAs

- Materials for print PSA campaign

  - Poster board

  - White paper

  - Colored construction paper

  - Markers

  - Scissors

  - Glue

SAFETY NOTES

1. Direct teacher supervision is imperative during all aspects of this activity to make sure students follow the safety guidelines.

2. All laboratory occupants must wear indirectly vented chemical splash goggles during all phases of this inquiry activity.

3. Immediately wipe up any water that is spilled on the floor to avoid a slip-and-fall hazard.

4. Use caution when working with sharps, which can cut or puncture skin.

5. Handle glassware and plasticware with care, as these can break and cut or puncture skin.

6. Any electrical power cords used near water must be plugged into a GFCI protected outlet. The teacher may need to provide a GFCI power strip for this use.

7. Use caution when working with glue guns, which get hot and can burn skin.

8. Wash hands with soap and water after completing the activity.

CONTENT STANDARDS AND KEY VOCABULARY

Table 4.10 lists the content standards from the *NGSS, CCSS,* and the Framework for 21st Century Learning that this lesson addresses, and Table 4.11 (p. 199) presents the key vocabulary. Vocabulary terms are provided for both teacher and student use. Teachers may choose to introduce some or all of the terms to students.

Table 4.10. Content Standards Addressed in STEM Road Map Module Lesson 4

NEXT GENERATION SCIENCE STANDARDS

PERFORMANCE EXPECTATIONS

- 5-ESS2-1. Develop a model using an example to describe ways in which the geosphere, biosphere, hydrosphere, and/or atmosphere interact.

- 5-ESS3-1. Obtain and combine information about ways individual communities use science ideas to protect the Earth's resources and environment.

- 5-LS2-1. Develop a model to describe the movement of matter among plants, animals, decomposers, and the environment.

SCIENCE AND ENGINEERING PRACTICES

Asking Questions and Defining Problems

- Ask questions that can be investigated and predict reasonable outcomes based on patterns such as cause and effect relationships.

Developing and Using Models

- Identify limitations of models.

- Collaboratively develop and/or revise a model based on evidence that shows the relationships among variables for frequent and regular occurring events.

- Develop and/or use models to describe and/or predict phenomena.

- Develop a diagram or simple physical prototype to convey a proposed object, tool, or process.

- Use a model to test cause and effect relationships or interactions concerning the functioning of a natural or designed system.

Continued

Table 4.10. (*continued*)

Planning and Carrying Out Investigations

- Plan and conduct an investigation collaboratively to produce data to serve as the basis for evidence, using fair tests in which variables are controlled and the number of trials considered.

- Evaluate appropriate methods and/or tools for collecting data.

- Make observations and/or measurements to produce data to serve as the basis for evidence for an explanation of a phenomenon or test a design solution.

- Make predictions about what would happen if a variable changes.

- Test two different models of the same proposed object, tool, or process to determine which better meets criteria for success.

Analyzing and Interpreting Data

- Represent data in tables and/or various graphical displays (bar graphs, pictographs, and/or pie charts) to reveal patterns that indicate relationships.

- Analyze and interpret data to make sense of phenomena, using logical reasoning, mathematics, and/or computation.

- Compare and contrast data collected by different groups in order to discuss similarities and differences in their findings.

- Analyze data to refine a problem statement or the design of a proposed object, tool, or process.

- Use data to evaluate and refine design solutions.

Using Mathematics and Computational Thinking

- Describe, measure, estimate, and/or graph quantities (e.g., area, volume, weight, time) to address scientific and engineering questions and problems.

- Create and/or use graphs and/or charts generated from simple algorithms to compare alternative solutions to an engineering problem.

Constructing Explanations and Designing Solutions

- Construct an explanation of observed relationships (e.g., the distribution of plants in the backyard).

- Use evidence (e.g., measurements, observations, patterns) to construct or support an explanation or design a solution to a problem.

- Identify the evidence that supports particular points in an explanation.

- Apply scientific ideas to solve design problems.

- Generate and compare multiple solutions to a problem based on how well they meet the criteria and constraints of the design.

Continued

Table 4.10. (*continued*)

Engaging in Argument From Evidence

- Compare and refine arguments based on an evaluation of the evidence presented.

- Respectfully provide and receive critiques from peers about a proposed procedure, explanation, or model by citing relevant evidence and posing specific questions.

- Construct and/or support an argument with evidence, data, and/or a model.

- Use data to evaluate claims about cause and effect.

- Make a claim about the merit of a solution to a problem by citing relevant evidence about how it meets the criteria and constraints of the problem.

Obtaining, Evaluating, and Communicating Information

- Obtain and combine information from books and other reliable media to explain phenomena.

- Read and comprehend grade-appropriate complex texts and/or other reliable media to summarize and obtain scientific and technical ideas and describe how they are supported by evidence.

- Communicate scientific and/or technical information orally and/or in written formats, including various forms of media as well as tables, diagrams, and charts.

DISCIPLINARY CORE IDEAS

ESS2.A: Earth Materials and Systems

- Earth's major systems are the geosphere (solid and molten rock, soil, and sediments), the hydrosphere (water and ice), the atmosphere (air), and the biosphere (living things, including humans). These systems interact in multiple ways to affect Earth's surface materials and processes. The ocean supports a variety of ecosystems and organisms, shapes landforms, and influences climate. Winds and clouds in the atmosphere interact with the landforms to determine patterns of weather.

ESS2.C: The Roles of Water in Earth's Surface Processes

- Nearly all of Earth's available water is in the ocean. Most fresh water is in glaciers or underground; only a tiny fraction is in streams, lakes, wetlands, and the atmosphere.

ESS3.C: Human Impacts on Earth Systems

- Human activities in agriculture, industry, and everyday life have had major effects on land, vegetation, streams, oceans, air, and even outer space. But individuals and communities are doing things to help protect Earth's resources and environments.

LS2.A: Interdependent Relationships in Ecosystems

- Decomposition eventually restores (recycles) some materials back to the soil. Organisms can survive only in environments in which their particular needs are met. A healthy ecosystem is one in which multiple species of different types are each able to meet their needs in a relatively stable web of life. Newly introduced species can damage the balance of an ecosystem.

Continued

Table 4.10. (*continued*)

LS2.B: Cycles of Matter and Energy Transfer in Ecosystems

- Matter cycles between air and soil and among plants, animals, and microbes as these organisms live and die. Organisms obtain gases and water from the environment and release waste matter (gas, liquid, or solid) back into the environment.

CROSSCUTTING CONCEPTS

Energy and Matter

- Energy can be transferred in various ways and between objects.

- Matter is transported into, out of, and within systems.

Scale, Proportion, and Quantity

- Standard units are used to measure and describe physical quantities such as weight and volume.

Systems and System Models

- A system can be described in terms of its components and their interactions.

COMMON CORE STATE STANDARDS FOR MATHEMATICS

MATHEMATICAL PRACTICES

- MP1. Make sense of problems and persevere in solving them.

- MP2. Reason abstractly and quantitatively.

- MP3. Construct viable arguments and critique the reasoning of others.

- MP4. Model with mathematics.

- MP5. Use appropriate tools strategically.

- MP6. Attend to precision.

- MP7. Look for and make use of structure.

MATHEMATICAL CONTENT

- 5.MD.A.1. Convert among different-sized standard measurement units within a given measurement system, and use these conversions in solving multi-step, real world problems.

- 5.MD.C.3. Recognize volume as an attribute of solid figures and understand concepts of volume measurement.

- 5.MD.C.3.A. A cube with side length 1 unit, called a "unit cube," is said to have "one cubic unit" of volume and can be used to measure volume.

- 5.MD.C.4. Measure volumes by counting unit cubes, using cubic cm, cubic inches, cubic feet, and improvised units.

- 5.MD.C.5. Relate volume to the operations of multiplication and addition and solve real world and mathematical problems using volume.

Continued

Table 4.10. (*continued*)

- 5.MD.C.5.B. Apply the formulas $V = l \times w \times h$ and $V = b \times h$ for rectangular prisms to find volumes of right rectangular prisms with whole-number edge lengths in the context of solving real world and mathematical problems.

- 5.NBT.A.3.A. Read and write decimals to thousandths using base-ten numerals and number names.

- 5.NBT.A.4. Use place value understanding to round decimals to any place.

- 5.NBT.B.5. Fluently multiply multi-digit whole numbers using the standard algorithm.

- 5.NBT.B.7. Add, subtract, multiply, and divide decimals to hundredths, using concrete models or drawings and strategies based on place value, properties of operations, and/or the relationship between addition and subtraction; relate the strategy to a written method and explain the reasoning used.

COMMON CORE STATE STANDARDS FOR ENGLISH LANGUAGE ARTS

READING STANDARDS

- RI.5.1. Quote accurately from a text when explaining what the text says explicitly and when drawing inferences from the text.

- RI.5.4. Determine the meaning of general and domain-specific words and phrases in a text relevant to a *grade 5 topic or subject area.*

- RI.5.7. Draw on information from multiple print or digital sources, demonstrating the ability to locate an answer to a question quickly or to solve a problem efficiently.

- RI.5.8. Explain how an author uses reasons and evidence to support particular points in a text, identifying which reasons and evidence support which point(s).

- RI.5.9. Integrate information from several texts on the same topic in order to write or speak about the subject knowledgeably.

- RF.5.3. Know and apply grade-level phonics and word analysis skills in decoding words.

- RF.5.4.A. Read grade-level text with purpose and understanding.

- RF.5.4.B. Read grade-level prose and poetry orally with accuracy, appropriate rate, and expression on successive readings.

WRITING STANDARDS

- W.5.2. Write informative/explanatory texts to examine a topic and convey ideas and information clearly.

- W.5.2.B. Develop the topic with facts, definitions, concrete details, quotations, or other information and examples related to the topic.

- W.5.2.C. Link ideas within and across categories of information using words, phrases, and clauses (e.g., *in contrast, especially*).

- W.5.2.D. Use precise language and domain-specific vocabulary to inform about or explain the topic.

Continued

Table 4.10. (*continued*)

- W.5.2.E. Provide a concluding statement or section related to the information or explanation presented.

- W.5.4. Produce clear and coherent writing in which the development and organization are appropriate to task, purpose, and audience.

- W.5.6. With some guidance and support from adults, use technology, including the internet, to produce and publish writing as well as to interact and collaborate with others; demonstrate sufficient command of keyboarding skills to type a minimum of two pages in a single sitting.

- W.5.7. Conduct short research projects that use several sources to build knowledge through investigation of different aspects of a topic.

- W.5.8. Recall relevant information from experiences or gather relevant information from print and digital sources; summarize or paraphrase information in notes and finished work, and provide a list of sources.

- W.5.9. Draw evidence from literary or informational texts to support analysis, reflection, and research.

SPEAKING AND LISTENING STANDARDS

- SL.5.1. Engage effectively in a range of collaborative discussions with diverse partners on *grade 5 topics and texts*, building on others' ideas and expressing their own clearly.

- SL.5.1.A. Come to discussions prepared, having read or studied required material; explicitly draw on that preparation and other information known about the topic to explore ideas under discussion.

- SL.5.1.B. Follow agreed-upon rules for discussions and carry out assigned roles.

- SL.5.1.C. Pose and respond to specific questions by making comments that contribute to the discussion and elaborate on the remarks of others.

- SL.5.1.D. Review key ideas expressed and draw conclusions in light of information and knowledge gained from the discussions.

- SL.5.4. Report on a topic or text or present an opinion, sequencing ideas logically and using appropriate facts and relevant, descriptive details to support main ideas or themes; speak clearly at an understandable pace.

- SL.5.5. Include multimedia components (e.g., graphics, sound) and visual displays in presentations when appropriate to enhance the development of main ideas or themes.

- SL.5.6. Adapt speech to a variety of contexts and tasks, using formal English when appropriate to task and situation.

FRAMEWORK FOR 21ST CENTURY LEARNING

- Interdisciplinary Themes: Health and Safety; Environmental Literacy; Science; Mathematics

- Learning and Innovation Skills: Creativity and Innovation; Critical Thinking and Problem Solving; Communication and Collaboration

Continued

Table 4.10. (*continued*)

- Information, Media, and Technology Skills: Information Literacy; Media Literacy
- Life and Career Skills: Flexibility and Adaptability; Initiative and Self-Direction; Social and Cross-Cultural Skills; Productivity and Accountability; Leadership and Responsibility

Table 4.11. Key Vocabulary for Lesson 4

Key Vocabulary	Definition
diameter	a straight line passing from one side to another through the center of a circle
media	ways to communicate information to large audiences, such as television, radio, the internet, newspapers, signs in public places, and other printed materials such as brochures
plumbing	a system of pipes, tanks, and fittings necessary for the water supply, heating, and sanitation of a building
pump	a mechanical device that uses pressure to move water through the plumbing system
spreadsheet	an electronic document in which data are arranged in columns and rows of a grid and can be manipulated and used to make calculations and graphs
tank	a large storage container used for liquid or gas substances

TEACHER BACKGROUND INFORMATION

The lesson begins with Angie's panicky e-mail introducing a new wrinkle in the Sunny Acres rainwater project. She has discovered that their plans to recycle rainwater off the parking lot have a couple of problems:

- The parking lot is nearly 3 meters below the area where the garden is located.
- A collection tank for the parking lot would have to be below the level of the parking lot for rainwater to flow into it.

The science activity, Acme Tank Works, focuses on finding a solution to this problem using a Cylinder Volume spreadsheet. To move forward with the project, teams must determine whether they should continue with the plan to harvest parking lot water or to consider capturing the rainwater that falls on the two buildings on either side of the garden.

That the parking lot is below the garden is not a major problem. The team could design a system with a pump that moves the water from a collection tank near the parking lot to a holding tank near the garden. In fact, if the collection tank can hold the water until the next sunny day, it may be possible to power the pump with solar panels. For teams that choose the parking lot scenario, the bigger problem is how to situate a tank large enough to capture the rainwater from the parking lot.

The second science activity assumes that the collection tank problem has been solved but introduces a new dilemma: the garden consists of two raised boxes. The top of each box is 75 cm above ground level. Once again, students are challenged to solve a problem that involves moving water from one place to another. This time, they must make a system that will split water flow into two separate streams. This may be accomplished with a tee fitting built from straws or tubing, although students may devise another solution to this problem. Figure 4.5 provides an example of a tee fitting that you may wish to show students who are struggling with the challenge of splitting the water flow into two streams.

Figure 4.5. Example of a Tee Fitting to Split Water Flow

Calculating Water Needs

In this lesson, students use mathematics to determine the water needed for the fictional garden in the challenge scenario. The following information may be useful to you:

- The data necessary to calculate the flow rate of the hose and the number of minutes needed to water the garden are provided in a comment found on Angie's Garden Sketch student handout: "Grandpa Henry says that the hose is slow. It takes 3 minutes to fill a 5 gallon bucket."

- Grandpa Henry waters for 20 minutes a day. The hose can deliver 5 gallons in 3 minutes, so the formula for total gallons is (5 gallons ÷ 3 minutes) × 20 minutes = 33.3 gallons. To convert daily water usage to liters, students should use the following formula:

    - Multiply → 33.3 gallons × 3.8 liters/gallon = 127 liters

- Once the teams determine the total daily water requirement, they will be able to calculate the number of days' water their tank can hold by using the following formula:

    - Multiply → the volume of their tank in liters × (1 day ÷ 127 liters/day)

The tank volume will vary by team according to the design the team arrived at in the Collection Tank Design Challenge. If we use the example that was cited in Angie's e-mail, to collect all the water from the playground would require a tank that could hold 3,750 liters. This tank would be able to provide about 30 days of water for the garden if Grandpa Henry waters for 20 minutes each day.

Using Spreadsheets

In this lesson, students write formulas in Microsoft Excel or Google Sheets to use the power of a computer spreadsheet to solve repetitive problems. In both Google Sheets and Microsoft Excel, a formula is begun in the cell by using an equals sign. Students then reference the values in cells by typing the cell reference into the formula. This exercise provides the height of a cylinder based on the diameter of the base and the target volume. A Google Sheets version of the Cylinder Volume spreadsheet is available at *https://tinyurl.com/AcmeCylinderTank*. The diameter row has a series of values for the tank diameter in centimeters. This is the independent variable. As the diameter increases, the height becomes shorter.

Pi and the target volume are constants in this activity. Reference a constant by adding dollar signs ($) as follows:

- Pi can be referenced by using B7 in the formula.

- Target volume can be referenced by using B6 in the formula.

Here are some other formulas and spreadsheet notations students will need to know:

- The formula for radius in meters uses the slash mark for division: = (B3/2)/100.

- The formula for area in square meters uses the asterisk for multiplication: = B4*B4*B7.

- The formula for height in cm uses area and target volume: = (B6/B5)*100.

Once the formulas are set up, they can be copied into the appropriate columns. Charts and graphs can be produced based on students' data by using commands within the program.

COMMON MISCONCEPTIONS

Students will have various types of prior knowledge about the concepts introduced in this lesson. Table 4.12 outlines some common misconceptions students may have concerning these concepts. Because of the breadth of students' experiences, it is not possible to anticipate every misconception that students may bring as they approach this lesson. Incorrect or inaccurate prior understanding of concepts can influence student learning in the future, however, so it is important to be alert to misconceptions such as those presented in the table.

Table 4.12. Common Misconceptions About the Concepts in Lesson 4

Topic	Student Misconception	Explanation
Models	Models are art projects.	Models are used to demonstrate and explain concepts that may be difficult to describe using only words. They may contain artistic elements, but their purpose is more than artistic.
	Models need to show every part of the object they represent.	Models should show the major features of the object they represent but do not need to include every detail.
	Models cannot be changed once they are constructed.	Using the EDP to construct a model means that the model can and should be changed and improved on so that it does a better job of demonstrating and explaining the function of the object it represents.

PREPARATION FOR LESSON 4

Review the Teacher Background Information (p. 199), assemble the materials for the lesson, and preview the videos recommended in the Learning Components section below.

To prepare for mathematics activities, reserve the computer lab for students to work on their Cylinder Volume spreadsheets. Download a copy of the spreadsheet from *http://tinyurl.com/AcmeCylinderTank.* You may either keep it on hand for your reference or furnish it to the class as a working model.

Student teams will create slideshow presentations of their challenge solutions. Create a template for the slideshow that includes an opening slide with the name of the challenge, "Rainwater Roundup," and be prepared to review the procedures for creating a slideshow if necessary. Reserve the computer lab for students to do additional research and to prepare their slideshows for the final project presentation. You may wish to invite parents and guardians, administrators, and the custodian or building manager to join the class for the final presentation of the teams' models and presentations. In your invitation, provide a brief overview of the Rainwater Roundup Challenge.

In ELA, each team will create materials for one component of the PSA campaign. Ensure that you have access to technology and materials for each type of media the class may choose. You may also incorporate a presentation of the class PSA components with the slideshow presentations teams create about their challenge solutions.

LEARNING COMPONENTS
Introductory Activity/Engagement

Connection to the Challenge: Begin each day of this lesson by directing students' attention to the driving question for the module: How can we use what we know about rainfall to design a system to provide water to a garden? Hold a brief student discussion of student prior knowledge on this topic, creating a class list of key ideas on chart paper, or you may wish to have students create a notebook entry with this information.

Driving Question for Lesson 4: How can we collect, store, and distribute rainwater to be used in a community garden?

Mathematics Class and Science Connection: Pass out copies of the Angie's E-mail With Playground Sketch handout to students. Tell students that you've received an urgent e-mail from Angie Zvatney at Wilbur Wright Elementary. Read the message from Angie's e-mail to the class. Ask students what problem Angie is facing (gravity). Review the components on the playground sketch on the handout as a class. Tell students that their challenge is to help their friends at Wilbur Wright find a solution for this problem. Then, distribute the Rainwater Roundup Challenge Checklist handout and the EDP Applied to the Rainwater Roundup handout to each student. As a class, review the handouts.

ELA and Social Studies Connections: Remind students that they have created a class message about watershed protection, referencing the Writing the OREO Way message displayed on chart paper. Review the message (the opinion) and the evidence the class chose to support this message via their PSA campaign. Remind students that in Lesson 3 each team was assigned a form of communication for which they will create materials.

Have students work together in teams to brainstorm their ideas for how to convey the message using their assigned form of communication.

Activity/Exploration

Mathematics Class and Science Connection: Introduce the following scenario to the class:

> *The folks from Sunny Acres called to ask if we can contact a local company called Acme Tank Works to have them make a plastic tank for their water storage tank. The people at Acme Tank Works said they can help, but they are in need of a quick way to calculate different tank sizes for their cylindrical plastic tanks. They make a whole range of sizes from 100 cm to 500 cm in diameter. When customers call to find a tank, they usually know how much water the tank should hold, but they often don't know how big the tank will be. They have asked us to provide information on the size of the tank we need, and they have also asked us to help them be able to find this information themselves in the future using a spreadsheet that can speed up their calculations.*

Acme Tank Works

Students should have access to the What's the Volume? handout from Lesson 3 for this activity. Tell students that they will use a spreadsheet (such as Google Sheets or Microsoft Excel) to provide a quick way for Acme Tank Works to calculate tank size. Pass out the Cylinder Volume Spreadsheet Example handout to each team, and review the example provided on *http://tinyurl.com/AcmeCylinderTank*.

Collection Tank Design Challenge

Next, remind teams about the message from Angie at Wilbur Wright Elementary and their ideas for the rainwater-recycling project. Lead students in a discussion about the problem by asking the following:

- What do we know about the layout of the Sunny Acres buildings, garden, and parking lot?

- What do we know about how water flows?

- Suppose you were planning to use the water from the parking lot. How could the water be collected?

- Is there any way to get the water to run uphill?

With what students know about Sunny Acres, there are two options for collecting the water: It can be collected from the building rooftops or at the low sides of the parking lot. Students have been told that the Sunny Acres parking lot is about the same size as their school playground. Distribute the Collection Tank Design Challenge Rubric (p. 217) to each team. Have each team choose one of the two options for collecting water and then have teams design a collection tank or tanks for the site they chose. The requirements for their designs are as follows:

- The tanks should fit in a footprint of 8 × 10 feet (about 2 ½ × 3 m) or less.

- The tanks must be positioned below the surface that receives the rainwater:

  - For the parking lot, tanks would need to be below the surface of the lot.

  - For the building rooftop, tanks would need to be below the lowest edge of the roof.

- Models of the tanks should be built out of 8 ounces of modeling clay in a 1:10 or 1:20 ratio, or to a round-number scale that makes sense. The finished models of the tanks should be about the size of a school milk carton.

- Display the models on the teams' Schoolyard Maps approximately where they would be located if the tanks were to be deployed at their school.

- Each model should include a notecard with data about the tank's dimensions and capacity in liters or cubic meters.

Garden Water Investigation

Remind students that the goal is to use the recycled rainwater to provide water for the garden. Angie's recent e-mail included interesting information about the garden's watering requirements. Grandpa Henry says he waters each side of the garden for 10 minutes each day, although he admits that the hose runs slowly. He states that it takes 3 minutes to fill a 5 gallon bucket.

Distribute the Angie's Garden Sketch and Garden Water Investigation student handouts to each team. Make sure that teams also have on hand the irrigation research they conducted in Lesson 3, as well as their tank volume calculations from the Collection Tank Design Challenge. Have teams use this information to calculate data that are necessary for the final design of their rainwater-recycling system. Teams should calculate the following:

- How many liters of water are necessary each day for the garden?

- Using the volume of your team's tank design, calculate the maximum number of days of water that your team's tank or tanks can hold.

- Using what you know about irrigation systems, recommend an alternative to using a standard garden hose and sprinkler.

STEM Research Notebook Prompt

Sketch an idea for a watering system showing your tanks and the raised garden. Sketch a close-up view of a section of the garden showing your suggestion for better irrigation.

Water Distribution Design Challenge

Tell students that you received another e-mail from Angie at Wilbur Wright Elementary School. This time she's not so panicked, because she knows the teams are working on the problem with her. In her e-mail, she informs you that the flower garden at Sunny Acres is in two raised boxes. The top of each box is 75 cm from the ground, and the boxes are separated by 1 meter so that residents can navigate between the garden rows in wheelchairs.

Teams are challenged to design and build a system that can distribute water from a single cup to two cups at a lower level. This system will be used to simulate watering the raised bed garden from the storage tank. The specifications for the system are as follows:

- The lower cups should be at least 0.5 m away from the supply cup.

- The supply cup needs to be raised only just enough above the destination cups that water will flow.

- The system should be able to be stopped and started.

- The system should distribute water evenly between the two destination cups.

- Teams may use the following materials: 3 paper or foam cups (12 oz.), 10 coffee stirrers, 10 flex straws, 10 paper clips, 20 craft sticks, 3 oz. modeling clay, glue, and duct tape in their designs.

STEM Research Notebook Prompt

Have students create a sketch of their team's water supply system and answer the following questions: *What was the most difficult part of this design challenge? If you were going to build a system like this to sell, what additional materials would make the job easier?*

Rainwater Roundup Challenge

Tell students that after having researched and created prototypes for the various parts of the rainwater-recycling system, the teams are ready to tackle the final challenge to design and build models of a rainwater-recycling system for the Sunny Acres Retirement Center. Answer any questions students have about the requirements listed on the Rainwater Roundup Challenge Checklist handout.

Students should use the checklist and the Applying the EDP to the Rainwater Roundup Challenge handout to guide their work. Each student should create a page in their STEM

Research Notebooks for each of the EDP steps to record their notes, rough sketches, and ideas. The following requirements should be addressed:

- The model should be built on a single foam board.

- The buildings and raised garden should be represented by boxes or suitable structures based on Angie's e-mails.

- Rainwater should be harvested from either the parking lot or building rooftops.

- Plumbing and irrigation lines should be shown from the tank to the garden beds.

- The ground around the raised beds should be free of hoses or other hazards that would make the area unsafe for Sunny Acres residents.

- The tanks and equipment should not detract from the beauty of the garden.

- The model should include notecards with technical information about the various portions of the system.

- If pumps are needed for the system, they should be represented on notecards.

- Teams may use the following materials: 1 foam board, 2 shoe boxes, 20 coffee stirrers, 20 flex straws, 1 m aquarium flexible tubing, 20 craft sticks, 10 rubber bands, 2 foam egg cartons, 3 oz. modeling clay, glue, and duct tape.

ELA and Social Studies Connections: Hand out the Public Service Advertising Campaign Rubric and review the expectations as a class. Each team should work together to plan and create its part of the PSA campaign about protecting the watershed. Provide each team with appropriate materials for the type of campaign it is creating, such as technology to make audio or video recordings and poster board, white paper, colored construction paper, markers, scissors, and glue for print media.

Explanation

Mathematics Class and Science Connection: Students should now plan and prepare a slideshow presentation of their proposal for a rainwater collection system for Sunny Acres staff and residents. Point out to students that an effective slide presentation does not include every word that the speaker says but rather is a summary of the major points. Use the Google slideshow on "Effective Presentations" at *http://tinyurl.com/Presentation-Coach* and the Rainwater Roundup Slideshow Rubric to explain the expectations for the presentation.

Give a copy of the Building an Argument handout to each team. Review with the class the components that should be in their presentations, emphasizing the following:

- *Identify the Problem.* The opening of the presentation should explain the problem students faced in the Rainwater Roundup Challenge by establishing the value of the garden and presenting their team's solution to the water problem. Choosing how to approach this problem was the first decision they made and should be introduced early in the presentation.

- *Choose the Best Solution.* Students should provide evidence to convince Sunny Acres staff and interested citizens that the team's solution fulfills their needs.

- *Plan the Argument.* An effective presentation begins with research and preparation. Explain the importance of supporting the team's decisions with evidence from scientific experimentation and research. Remind the teams about all the design decisions they made on the pathway to their solution:

  - The source for their rainwater: parking lot or roof

  - The size and placement of their storage tanks

  - The design of the system to get water from the tank or tanks to the raised beds

  - The irrigation design

Next, pass out a piece of copy paper to each student. Ask students to fold the paper in half three times so that there are eight rectangles when opened up. These eight rectangles will represent eight slides they will design. Tell students that the first slide must include the title of the project, Rainwater Roundup, and the team name. Emphasize to students that teamwork is important in creating the slideshow. Together they will decide which information is the most important to support their argument and keep their audience interested.

ELA and Social Studies Connections: Students should work on completing their PSA campaign materials. Teams that need to record their presentations (e.g., "television" and "radio" advertisements) should practice and then record their presentations so that they can be presented in the Elaboration/Application of Knowledge section.

Elaboration/Application of Knowledge

Mathematics Class and Science, Social Studies, and ELA Connections: Hold a presentation day for students' Rainwater Roundup Slideshow presentations. Encourage the audience (guests and other student teams) to ask questions about teams' designs and the design process they used. You may also wish to have students present their PSA campaign materials.

Have students decide how to distribute their PSA materials to the community. For example, they may wish to contact the local library or a local government office to ask for permission to display the materials there.

Evaluation/Assessment

Students may be assessed on the following performance tasks and other measures listed.

Performance Tasks

- Cylinder Volume spreadsheet tables and graphs
- Collection Tank Design Challenge Rubric
- Garden Water Investigation
- Water Distribution Design Challenge Rubric
- Rainwater Roundup Challenge Rubric
- Rainwater Roundup Slideshow Rubric
- Public Service Advertising Campaign Rubric

Other Measures

- STEM Research Notebook entries
- Engagement in class activities and discussions
- Involvement in group work and discussions

INTERNET RESOURCES

Cylinder Volume spreadsheet example
- *http://tinyurl.com/AcmeCylinderTank*

"Effective Presentations"
- *http://tinyurl.com/PresentationCoach*

IMAGE CREDITS FOR LESSON 4

Hose (p. 212)
www.flickr.com/photos/kpaulus/8517559208
Owner: Kristine Paulus; License: Attribution 2.0 Generic (CC BY 2.0)

Drawings and tables (Owner: Pandaia Projects LLC. Used with permission.)
- Angie's sketches (pp. 210 and 212)
- Cylinder volume table (p. 211)

STUDENT HANDOUT

ANGIE'S E-MAIL WITH PLAYGROUND SKETCH

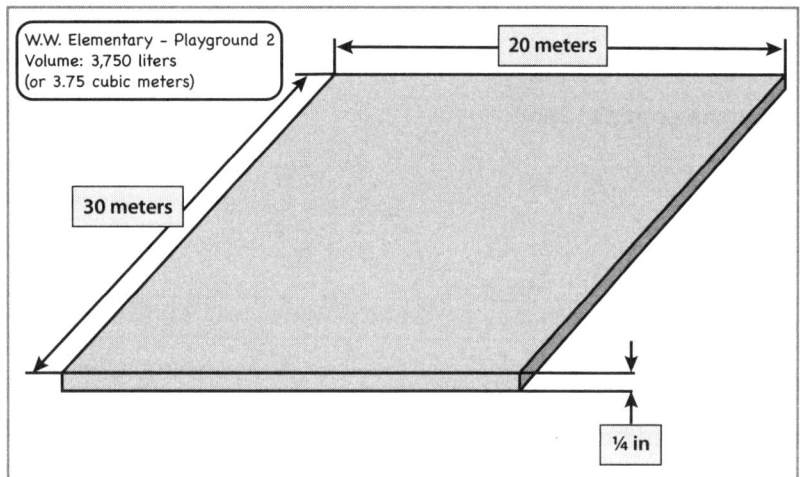

W.W. Elementary - Playground 2
Volume: 3,750 liters
(or 3.75 cubic meters)

20 meters

30 meters

¼ in

The playground is not very big, but it gets a lot of water! One morning's rain is over 3 cubic meters!

Jorge and I walked down to the playground, and we noticed that the playground is downhill from the school, just like the parking lot at Sunny Acres.

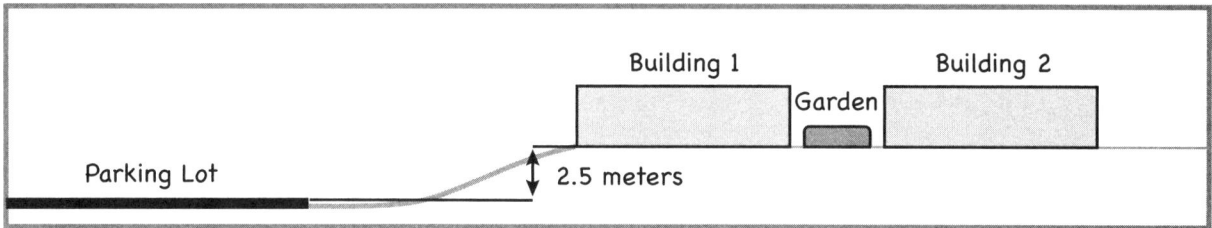

Building 1

Building 2

Garden

Parking Lot

2.5 meters

Sunny Acres Layout

OH No!!!

I just realized, the water on the parking lot is downhill from the garden! How are we going to get the water we collect up to the garden?

Jorge just reminded me of another OH NO!

How are we going to collect water from the parking lot? Sure, there's lots of water ... 3,750 liters for one little old 1/4" of rain. But it all goes down a storm drain!

When the team designed a tank that could hold that much water, it was pretty big - way too big to fit down in the ditch where the storm drain empties!

CYLINDER VOLUME SPREADSHEET EXAMPLE

A fully functional Google spreadsheet provided at *http://tinyurl.com/AcmeCylinderTank* can be used to determine the best size for the volume of the water tank based on its diameter versus height. You can use this spreadsheet or create your own. To use the provided spreadsheet, you only need to adjust the diameter for the graph to automatically be updated.

HO4-2 Cylinder Volume

CylinderTank | Playground

Cylinder Tank Builder

Height	478	395	332	283	244	212	187	165	147	132	119
Diameter	100	110	120	130	140	150	160	170	180	190	200
Radius in Meters	0.5	0.55	0.6	0.65	0.7	0.75	0.8	0.85	0.9	0.95	1
Area in Square Meters	0.79	0.95	1.13	1.33	1.54	1.77	2.01	2.27	2.54	2.83	3.14
Target Volume	3.75										
Pi	3.14										

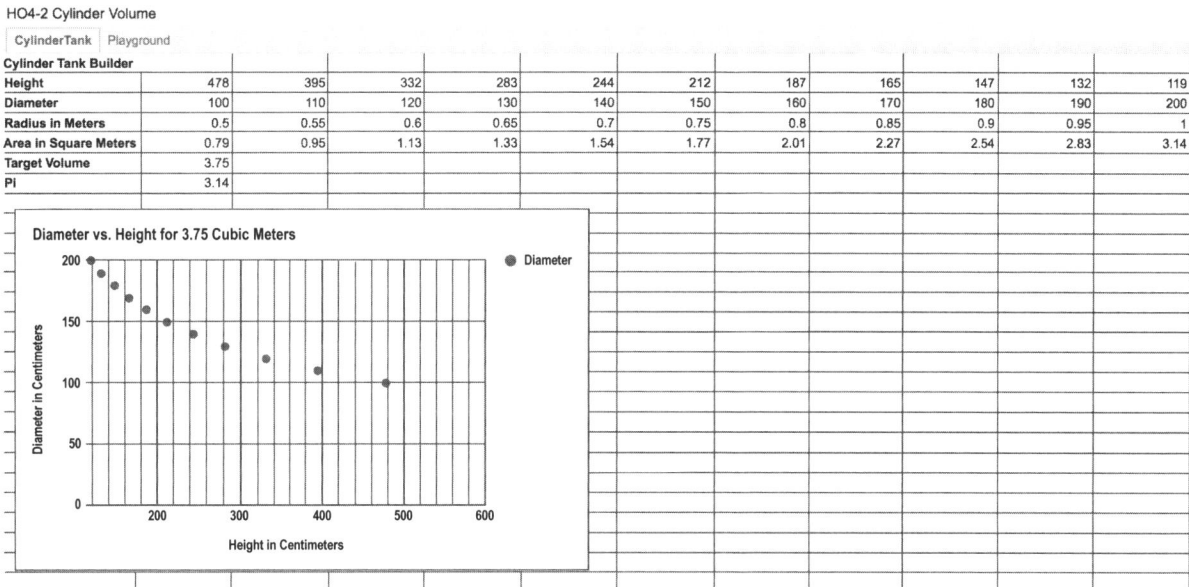

STUDENT HANDOUT

ANGIE'S GARDEN SKETCH

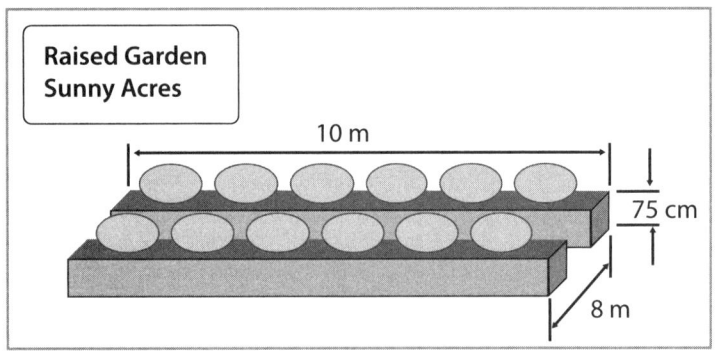

**Raised Garden
Sunny Acres**

10 m

75 cm

8 m

The garden at Sunny Acres is raised so that the residents don't have to bend over to take care of the plants.

Our watering system will need to water both sides.

Grandpa Henry says he uses a 5/8" hose to water the garden. He waters each side for 10 minutes every day

I wonder how much water the garden needs.

Grandpa Henry says that the hose is slow. It takes 3 minutes to fill a 5 gallon bucket with water.

Maybe we could use a different kind of sprinkler or something.

Team Name: _____   Date: _____

STUDENT HANDOUT

GARDEN WATER INVESTIGATION

Your team has been assigned to calculate water requirements for the Sunny Acres raised garden. How much water does the garden need? How much can our system supply?

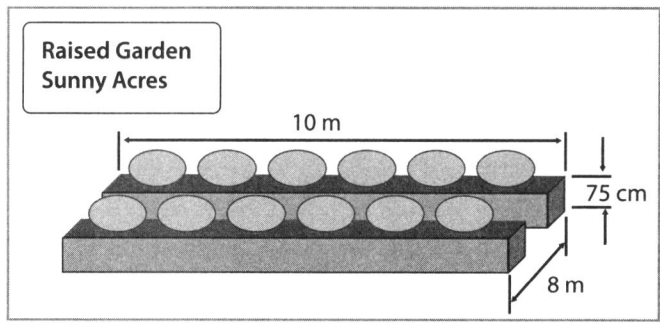

Raised Garden
Sunny Acres

10 m

75 cm

8 m

This week, the class received an e-mail from Angie at Wilbur Wright Elementary. She measured the garden at Sunny Acres and made some important discoveries. The garden is 8 m x 10 m, and the plants are in raised boxes, with the top of each box 75 cm off the ground. Angie interviewed Grandpa Henry about watering the garden. He waters each side of the garden for 10 minutes a day. He told Angie that the hose is so slow, it takes 3 minutes to fill a 5 gallon bucket to wash his tools.

Calculate the number of liters needed each day to water the garden. Round to the nearest whole liter:
Calculate the number of days' supply of water that your tanks can hold:
Consult your irrigation research from Lesson 3. Is there a better way to water than using a hose and sprinkler?
Why would a different irrigation method be better?

STUDENT HANDOUT

RAINWATER ROUNDUP CHALLENGE CHECKLIST

Now that you have researched rainwater-recycling systems and created prototypes for a rainwater collection tank and water distribution system, you will create a model of the rainwater-recycling system for the Sunny Acres Retirement Center.

Your team's model must meet the following requirements:

✓ The model should be built on a single foam board.

✓ The buildings and raised garden should be represented by boxes or suitable structures based on Angie's e-mails.

✓ Rainwater should be harvested from either the parking lot or building rooftops.

✓ Plumbing and irrigation lines should be shown from the tank to the garden beds.

✓ The ground around the raised beds should be free of hoses or other hazards that would make the area unsafe for Sunny Acres residents.

✓ The tanks and equipment should not detract from the beauty of the garden.

✓ The model should include notecards with technical information about the various portions of the system.

✓ If pumps are needed for the system, they should be represented on notecards.

✓ Your team must use the EDP, and all team members should create a page for each step of the EDP in their STEM Research Notebooks and record their notes, ideas, and rough sketches for each step of the EDP.

✓ Teams may use the following materials: 1 foam board, 2 shoe boxes, 20 coffee stirrers, 20 flex straws, 1 m aquarium flexible tubing, 20 craft sticks, 10 rubber bands, 2 foam egg cartons, 3 oz. modeling clay, glue, and duct tape.

STUDENT HANDOUT

EDP APPLIED TO THE RAINWATER ROUNDUP

Problem

Build a rainwater-recycling system model for Sunny Acres retirement center. It should collect water from either parking lots or building roofs. It should connect to an irrigation system for the center's raised-bed garden.

Task 1. Your team has been given some common craft materials. Perhaps you have built something like this with a 3-D building set. Brainstorm how you can use your materials to build a model of the system.

Task 2. The three main sections of this system are the collection system, the tanks, and the irrigation system. Plan how to build each section so that they work together.

Task 3. Get to work building your model using the plan you made. Be sure that the area around the raised beds is clear of hazards that could endanger the residents.

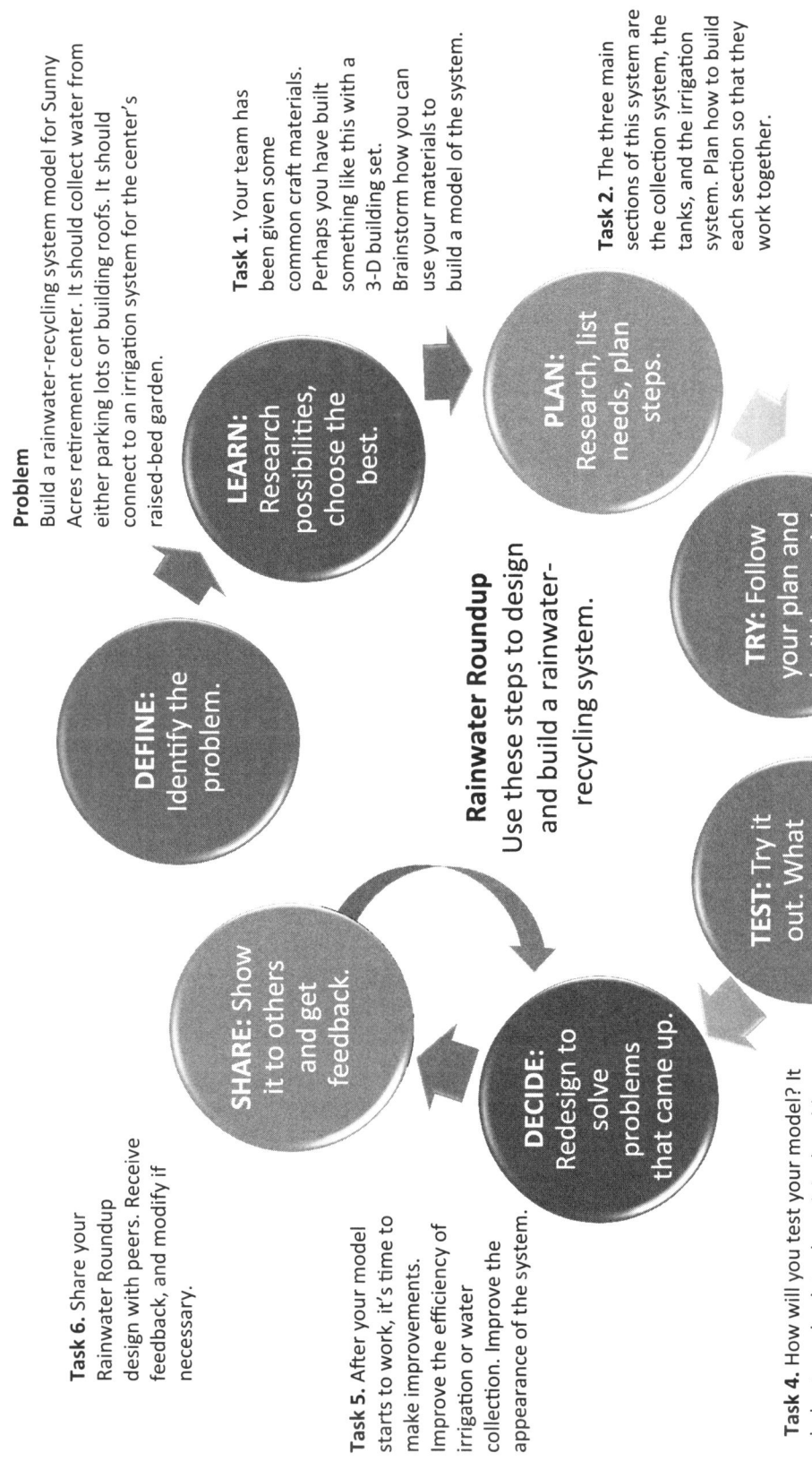

DEFINE: Identify the problem.

LEARN: Research possibilities, choose the best.

PLAN: Research, list needs, plan steps.

TRY: Follow your plan and build a model.

TEST: Try it out. What works? What doesn't?

DECIDE: Redesign to solve problems that came up.

SHARE: Show it to others and get feedback.

Rainwater Roundup
Use these steps to design and build a rainwater-recycling system.

Task 6. Share your Rainwater Roundup design with peers. Receive feedback, and modify if necessary.

Task 5. After your model starts to work, it's time to make improvements. Improve the efficiency of irrigation or water collection. Improve the appearance of the system.

Task 4. How will you test your model? It helps to go back to the original problem you are trying to solve. Decide on criteria. How many days' water will your model be able to supply?

STUDENT HANDOUT

BUILDING AN ARGUMENT

OUR SOLUTION TO THE PROBLEM …

Our Rainwater Collection system will solve the problem in an efficient and effective manner.

PRO ——— CON

Possible Questions

Others might ask …

How we would respond to these questions

Our Reasons

1.

2.

3.

Evidence to back up our reasons

Team Name: _____

Collection Tank Design Challenge Rubric

Criteria	Meets or Exceeds Expectations (5–6 points)	Approaches Expectations (3–4 points)	Needs Improvement (1–2 points)	Score
DEFINE	• Team shows clear understanding of problem. • Team shows clear understanding of constraints.	• Team shows clear understanding of problem. • Team considered constraints.	• Team shows incomplete understanding of problem. • Team does not consider constraints.	
LEARN	• Team clearly brainstormed multiple possible solutions. • Solution chosen from analysis.	• Team brainstormed around single solution. • Solution chosen before brainstorming.	• Little or no brainstorming. • Solution does not solve problem.	
PLAN	• Tank design includes a drawing. • Drawing is labeled with real-world dimensions. • Drawing is very neat and easy to understand.	• Tank design includes a drawing. • Drawing is labeled with real-world dimensions.	• Tank design does not include a drawing. OR • Drawing is provided but has few or no dimensions or notes.	
TRY	• Model closely follows drawings and is carefully and neatly constructed. • Notecard lists required information and additional details.	• Model resembles drawings. • Notecard lists required information.	• Model does not resemble drawing. • Notecard data omitted.	
TEST	• Model is appropriate to collection site (roof or parking lot), and it is clear how it would work in this space. • Model fits inside 8 × 10 foot footprint.	• Model may be appropriate to the collection site, but it is not clear how it would work in this space. • Model fits inside 8 × 10 foot footprint.	• Model is not appropriate for the collection site. • Fit in 8 × 10 foot footprint cannot be determined because of missing dimensions.	
DECIDE	• Clear evidence that the team made revisions and modifications to improve the design.	• Some evidence that the team made revisions and modifications to improve the design.	• No evidence that the team made revisions and modifications to improve the design.	

TOTAL SCORE: _____

COMMENTS:

Water Distribution Design Challenge Rubric

Team Name: _____

Criteria	Exceeds Expectations (4 points)	Meets Expectations (3 points)	Approaches Expectations (2 points)	Needs Improvement (1 point)	Score
DESIGN	• Design drawings created. • Precise dimensions present on drawings. • Includes notes about construction.	• Design drawings created. • Some dimensions present on drawings. • Includes notes about construction.	• Some design drawings present. • Some dimensions present on drawings. • Few notes to aid with construction.	• No design drawings were made. • No dimensions were considered. • No instructions or notes provided.	
BUILD	• Model closely resembles drawings. • Team members understand each component's function. • Model construction has finished appearance.	• Model closely resembles drawings. • Team members understand most component's functions. • Model is rough in appearance.	• Model somewhat resembles drawings. • Some team members understand some components' functions. • Model has a very rough appearance.	• Model does not resemble drawings. • Team members do not understand the functions of components. • Model is not complete.	
FUNCTION	• Minimal leakage. • Nearly all water in first cup drains to secondary cups. • Stop and start valve works.	• Some leakage. • Most water in first cup drains to secondary cups. • Stop and start valve works.	• Some leakage. • About half of water in first cup drains to secondary cups. • Valve does not work.	• Substantial leakage. • Less than half of water in first cup drains to secondary cups. • No valve.	
APPLICATION	• Model is clearly a prototype for garden. • Secondary cups separated from first cup by at least 0.5 meters. • Evidence of design improvements.	• Model is clearly a prototype for garden. • Secondary cups separated from first cup by at least 0.5 meters. • Some evidence of design improvements.	• Model could be prototype for garden. • Secondary cups separated from first cup by at least 0.25 meters. • Little evidence of design improvements.	• Model does not resemble garden layout. • Secondary cups separated from first cup by less than 0.25 meters. • No evidence of design improvements.	

TOTAL SCORE: _____

COMMENTS:

Rainwater Roundup Challenge Rubric

Team Name: _____

Criteria	Exceeds or Meets Expectations (5–6 points)	Approaches Expectations (3–4 points)	Needs Improvement (1–2 points)	Score
DEFINE	• Team shows clear understanding of problem. • Team shows clear understanding of limits.	• Team shows clear understanding of problem. • Team considered some limits.	• Team shows incomplete understanding of problem. • Team does not consider limits.	
LEARN	• Team brainstormed multiple possible solutions. • Solution chosen after analysis.	• Team brainstormed around single solution. • Solution chosen before analysis.	• Team did little or no brainstorming. • Solution does not solve problem.	
PLAN	• Design includes drawings for three sections: tanks, collection, irrigation. • Sketch includes details with notes that provide justification of design.	• Design includes drawings for two sections. • Few details are shown in sketch, and notes define but do not justify design.	• Design does not include a drawing. OR • Only a rough sketch with few notes is provided.	
TRY	• Model closely follows sketch. • Model closely resembles Angie's sketches. • Model includes detailed notecards.	• Model somewhat resembles sketch. • Model somewhat resembles Angie's sketches. • Model includes notecards but little detail.	• Model resembles sketch. • Model does not resemble Angie's sketches. • Vital data omitted on notecards.	
TEST	• Model is tailored to Sunny Acres. • Model applies physics laws, and they are well documented in notes. • Model design considers both safety AND appearance.	• Model is appropriate to intended site. • Model may work, but physics laws are only loosely applied. • Model design considers safety OR appearance.	• Model not tailored for site. • Model would violate physics laws. • Model design considers neither safety NOR appearance.	
DECIDE	• Evidence of correction based on feedback from the model.	• Evidence of improvement based on feedback.	• No evidence of alteration after feedback.	

TOTAL SCORE: _____

COMMENTS:

Rainwater Roundup Slideshow Rubric

Team Name: _____

Criteria	Exceeds Expectations (5–6 points)	Meets Expectations (3–4 points)	Needs Improvement (1–2 points)	Score
INFORMATION	• Team presentation includes many relevant details with extra pieces of information about the problem. • Provides an abundance of evidence to clearly support the reasoning.	• Team presentation includes many relevant details to demonstrate an understanding of the problem. • Proposal is made with evidence to support the reasoning.	• Team presentation omits many important details. • Proposal is made without evidence to support the reasoning.	
ACCURACY	• All the content is correct. • Understanding of content is obvious.	• Most of the content is correct. • Some gaps in understanding are evident.	• Much of the content is not correct. • There is little evidence of understanding of content.	
TIME MANAGEMENT	• Manages time well. • Presentation is clear, well organized, and informative.	• Manages time adequately. • Presentation is semi-organized and informative.	• Manages time poorly. Team is frequently off task. • Presentation is difficult to follow.	
GRAPHICS	• Images are clear. • Images are used effectively and can be used to support conclusions.	• Images are clear. • Some images are helpful but may not support conclusions.	• Images are unclear. • Images distract from the presentation.	
MECHANICS	• The project has no more than one spelling error. • The project has no more than one grammatical error.	• The project has two to four spelling errors. • The project has two to four grammatical errors.	• The project has five or more spelling errors. • The project has five or more grammatical errors.	
Q&A RESPONSES	• Team responds to and seeks clarification of audience questions. • Responses are full of details that clarify.	• Team responds to audience questions. • Responses are brief and incomplete.	• Team fails to respond to questions from audience. • Team is not able to answer many of the questions.	

TOTAL SCORE: _____

COMMENTS:

Public Service Advertising Campaign Rubric

Team Name: _____

Criteria	Exceeds Expectations (4 points)	Meets Expectations (3 points)	Approaches Expectations (2 points)	Needs Improvement (1 point)	Score
INFORMATION	Team included many interesting facts and covered most of the key ideas of the topic(s).	Team included some interesting facts and introduced many key ideas of the topic(s).	Team included a few interesting facts and omitted several key ideas of the topic.	Team presentation was not interesting and omitted many key ideas of the topic.	
ORGANIZATION	There was a clear organizational strategy, and all information was presented logically in a way that enhances reader or viewer understanding.	There was an organizational strategy, and most information was presented logically.	There was an organizational strategy, but the information was not presented logically and did not enhance reader or viewer understanding.	There was not a clear organizational strategy, and the information was not presented logically.	
CONTENT	The product clearly conveyed the message the class decided on and incorporated all reasons and examples in meaningful and creative ways.	The product conveyed the message the class decided on and incorporated all reasons and examples.	The product included the message the class decided on but did not incorporate all reasons and examples.	The message of the product was unclear, and few reasons and examples were used.	
DELIVERY	The product was polished, neat, and visually appealing. If it included speech, the speech was clear and easy to understand.	The product was neat and contained visually appealing elements. If it included speech, the speech was clear and easy to understand.	The product was neat but not visually appealing. If it included speech, the speech was difficult to understand at times.	The product was not neat or visually appealing. If it included speech, the speech was difficult to understand.	

TOTAL SCORE: _____

COMMENTS:

TRANSFORMING LEARNING WITH RAINWATER ANALYSIS AND THE *STEM ROAD MAP CURRICULUM SERIES*

Carla C. Johnson

This chapter serves as a conclusion to the Rainwater Analysis integrated STEM curriculum module, but it is just the beginning of the transformation of your classroom that is possible through use of the *STEM Road Map Curriculum Series.* In this book, many key resources have been provided to make learning meaningful for your students through integration of science, technology, engineering, and mathematics, as well as social studies and English language arts, into powerful problem- and project-based instruction. First, the Rainwater Analysis curriculum is grounded in the latest theory of learning for students in grade 5 specifically. Second, as your students work through this module, they engage in using the engineering design process (EDP) and build prototypes like engineers and STEM professionals in the real world. Third, students acquire important knowledge and skills grounded in national academic standards in mathematics, English language arts, science, and 21st century skills that will enable their learning to be deeper, retained longer, and applied throughout, illustrating the critical connections within and across disciplines. Finally, authentic formative assessments, including strategies for differentiation and addressing misconceptions, are embedded within the curriculum activities.

The Rainwater Analysis curriculum in The Represented World STEM Road Map theme can be used in single-content classrooms where there is only one teacher or expanded to include multiple teachers and content areas across classrooms. Through the exploration of the Rainwater Roundup Challenge, students engage in a real-world STEM problem on the first day of instruction and gather necessary knowledge and skills along the way in the context of solving the problem.

The other topics in the *STEM Road Map Curriculum Series* are designed in a similar manner, and NSTA Press has additional volumes in this series for this and other grade levels and plans to publish more. The volumes covering Innovation and Progress have been published and are as follows:

- *Amusement Park of the Future, Grade 6*

- *Construction Materials, Grade 11*

- *Harnessing Solar Energy, Grade 4*

- *Transportation in the Future, Grade 3*

- *Wind Energy, Grade 5*

In addition, several volumes covering The Represented World have been published:

- *Car Crashes, Grade 12*

- *Improving Bridge Design, Grade 8*

- *Packaging Design, Grade 6*

- *Patterns and the Plant World, Grade 1*

- *Swing Set Makeover, Grade 3*

The tentative list of other books includes the following themes and subjects:

- The Represented World (*continued*)

  - Investigating environmental changes

  - Radioactivity

- Cause and Effect

  - Influence of waves

  - Hazards and the changing environment

  - The role of physics in motion

- Sustainable Systems

  - Creating global bonds

  - Composting: Reduce, reuse, recycle

  - Hydropower efficiency

  - System interactions

- Optimizing the Human Experience

  - Genetically modified organisms

  - Mineral resources

  - Rebuilding the natural environment

  - Water conservation: Think global, act local

If you are interested in professional development opportunities focused on the STEM Road Map specifically or integrated STEM or STEM programs and schools overall, contact the lead editor of this project, Dr. Carla C. Johnson (*carlacjohnson@purdue.edu*), associate dean and professor of science education at Purdue University. Someone from the team will be in touch to design a program that will meet your individual, school, or district needs.

APPENDIX

CONTENT STANDARDS ADDRESSED IN THIS MODULE

NEXT GENERATION SCIENCE STANDARDS

Table A1 (p. 228) lists the science and engineering practices, disciplinary core ideas, and crosscutting concepts this module addresses. The supported performance expectations are as follows:

- 5-ESS2-1. Develop a model using an example to describe ways in which the geosphere, biosphere, hydrosphere, and/or atmosphere interact.

- 5-ESS2-2. Describe and graph the amounts and percentages of water and fresh water in various reservoirs to provide evidence about the distribution of water on Earth.

- 5-ESS3-1. Obtain and combine information about ways individual communities use science ideas to protect the Earth's resources and environment.

- 5-LS2-1. Develop a model to describe the movement of matter among plants, animals, decomposers, and the environment.

Table A1. *Next Generation Science Standards (NGSS)*

Science and Engineering Practices

ASKING QUESTIONS AND DEFINING PROBLEMS

- Ask questions that can be investigated and predict reasonable outcomes based on patterns such as cause and effect relationships.

DEVELOPING AND USING MODELS

- Identify limitations of models.

- Collaboratively develop and/or revise a model based on evidence that shows the relationships among variables for frequent and regular occurring events.

- Develop a model using an analogy, example, or abstract representation to describe a scientific principle or design solution.

- Develop and/or use models to describe and/or predict phenomena.

- Develop a diagram or simple physical prototype to convey a proposed object, tool, or process.

- Use a model to test cause and effect relationships or interactions concerning the functioning of a natural or designed system.

PLANNING AND CARRYING OUT INVESTIGATIONS

- Plan and conduct an investigation collaboratively to produce data to serve as the basis for evidence, using fair tests in which variables are controlled and the number of trials considered.

- Evaluate appropriate methods and/or tools for collecting data.

- Make observations and/or measurements to produce data to serve as the basis for evidence for an explanation of a phenomenon or test a design solution.

- Make predictions about what would happen if a variable changes.

- Test two different models of the same proposed object, tool, or process to determine which better meets criteria for success.

ANALYZING AND INTERPRETING DATA

- Represent data in tables and/or various graphical displays (bar graphs, pictographs, and/or pie charts) to reveal patterns that indicate relationships.

- Analyze and interpret data to make sense of phenomena, using logical reasoning, mathematics, and/or computation.

- Compare and contrast data collected by different groups in order to discuss similarities and differences in their findings.

- Analyze data to refine a problem statement or the design of a proposed object, tool, or process.

- Use data to evaluate and refine design solutions.

Continued

Table A1. (*continued*)

USING MATHEMATICS AND COMPUTATIONAL THINKING

- Organize simple data sets to reveal patterns that suggest relationships.

- Describe, measure, estimate, and/or graph quantities (e.g., area, volume, weight, time) to address scientific and engineering questions and problems.

- Create and/or use graphs and/or charts generated from simple algorithms to compare alternative solutions to an engineering problem.

CONSTRUCTING EXPLANATIONS AND DESIGNING SOLUTIONS

- Construct an explanation of observed relationships (e.g., the distribution of plants in the backyard).

- Use evidence (e.g., measurements, observations, patterns) to construct or support an explanation or design a solution to a problem.

- Identify the evidence that supports particular points in an explanation.

- Apply scientific ideas to solve design problems.

- Generate and compare multiple solutions to a problem based on how well they meet the criteria and constraints of the design.

ENGAGING IN ARGUMENT FROM EVIDENCE

- Compare and refine arguments based on an evaluation of the evidence presented.

- Respectfully provide and receive critiques from peers about a proposed procedure, explanation, or model by citing relevant evidence and posing specific questions.

- Construct and/or support an argument with evidence, data, and/or a model.

- Use data to evaluate claims about cause and effect.

- Make a claim about the merit of a solution to a problem by citing relevant evidence about how it meets the criteria and constraints of the problem.

OBTAINING, EVALUATING, AND COMMUNICATING INFORMATION

- Obtain and combine information from books and/or other reliable media to explain phenomena.

- Read and comprehend grade-appropriate complex texts and/or other reliable media to summarize and obtain scientific and technical ideas and describe how they are supported by evidence.

- Combine information in written text with that contained in corresponding tables, diagrams, and/or charts to support the engagement in other scientific and/or engineering practices.

- Communicate scientific and/or technical information orally and/or in written formats, including various forms of media as well as tables, diagrams, and charts.

Continued

Table A1. (*continued*)

Disciplinary Core Ideas

ESS2.A: EARTH MATERIALS AND SYSTEMS

- Earth's major systems are the geosphere (solid and molten rock, soil, and sediments), the hydrosphere (water and ice), the atmosphere (air), and the biosphere (living things, including humans). These systems interact in multiple ways to affect Earth's surface materials and processes. The ocean supports a variety of ecosystems and organisms, shapes landforms, and influences climate. Winds and clouds in the atmosphere interact with the landforms to determine patterns of weather.

ESS2.C: THE ROLES OF WATER IN EARTH'S SURFACE PROCESSES

- Nearly all of Earth's available water is in the ocean. Most fresh water is in glaciers or underground; only a tiny fraction is in streams, lakes, wetlands, and the atmosphere.

ESS3.C: HUMAN IMPACTS ON EARTH SYSTEMS

- Human activities in agriculture, industry, and everyday life have had major effects on land, vegetation, streams, oceans, air, and even outer space. But individuals and communities are doing things to help protect Earth's resources and environments.

LS1.C: ORGANIZATION FOR MATTER AND ENERGY FLOW IN ORGANISMS

- Plants acquire their material for growth chiefly from air and water.

LS2.A: INTERDEPENDENT RELATIONSHIPS IN ECOSYSTEMS

- Decomposition eventually restores (recycles) some materials back to the soil. Organisms can survive only in environments in which their particular needs are met. A healthy ecosystem is one in which multiple species of different types are each able to meet their needs in a relatively stable web of life. Newly introduced species can damage the balance of an ecosystem.

LS2.B: CYCLES OF MATTER AND ENERGY TRANSFER IN ECOSYSTEMS

- Matter cycles between air and soil and among plants, animals, and microbes as these organisms live and die. Organisms obtain gases and water from the environment and release waste matter (gas, liquid, or solid) back into the environment.

Crosscutting Concepts

ENERGY AND MATTER

- Energy can be transferred in various ways and between objects.
- Matter is transported into, out of, and within systems.

SCALE, PROPORTION, AND QUANTITY

- Standard units are used to measure and describe physical quantities such as weight and volume.

SYSTEMS AND SYSTEM MODELS

- A system can be described in terms of its components and their interactions.

Source: NGSS Lead States. 2013. *Next Generation Science Standards: For states, by states.* Washington, DC: National Academies Press. *www.nextgenscience.org/next-generation-science-standards.*

Table A2. Common Core Mathematics and English Language Arts (ELA) Standards

MATHEMATICAL PRACTICES

- MP1. Make sense of problems and persevere in solving them.
- MP2. Reason abstractly and quantitatively.
- MP3. Construct viable arguments and critique the reasoning of others.
- MP4. Model with mathematics.
- MP5. Use appropriate tools strategically.
- MP6. Attend to precision.
- MP7. Look for and make use of structure.

MATHEMATICAL CONTENT

- 5.GA.1. Use a pair of perpendicular number lines, called axes, to define a coordinate system, with the intersection of lines (the origin) arranged to coincide with the 0 on each line and a given point in the plane located by using an ordered pair of numbers, called its coordinates. Understand that the first number indicates how far to travel from the origin in the direction of one axis, and the second number indicates how far to travel in the direction of the second axis, with the convention that the names of the two axes and the coordinates correspond (e.g., x-axis and x-coordinate, y-axis and y-coordinate).
- 5.GA.2. Represent real world and mathematical problems by graphing points in the first quadrant of the coordinate plane, and interpret coordinate values of points in the context of the situation.
- 5.MD.A.1. Convert among different-sized standard measurement units within a given measurement system, and use these conversions in solving multi-step, real world problems.
- 5.MD.C.3. Recognize volume as an attribute of solid figures and understand concepts of volume measurement.
- 5.MD.C.3.A. A cube with side length 1 unit, called a "unit cube," is said to have "one cubic unit" of volume and can be used to measure volume.
- 5.MD.C.4. Measure volumes by counting unit cubes, using cubic cm, cubic inches, cubic feet, and improvised units.

READING STANDARDS

- RI.5.1. Quote accurately from a text when explaining what the text says explicitly and when drawing inferences from the text.
- RI.5.4. Determine the meaning of general and domain-specific words and phrases in a text relevant to a *grade 5 topic or subject area.*
- RI.5.7. Draw on information from multiple print or digital sources, demonstrating the ability to locate an answer to a question quickly or to solve a problem efficiently.
- RI.5.8. Explain how an author uses reasons and evidence to support particular points in a text, identifying which reasons and evidence support which point(s).
- RI.5.9. Integrate information from several texts on the same topic in order to write or speak about the subject knowledgeably.
- RF.5.3. Know and apply grade-level phonics and word analysis skills in decoding words.
- RF.5.4.A. Read grade-level text with purpose and understanding.
- RF.5.4.B. Read grade-level prose and poetry orally with accuracy, appropriate rate, and expression on successive readings.

WRITING STANDARDS

- W.5.1. Write opinion pieces on topics or texts, supporting a point-of-view with reasons and information.
- W.5.2. Write informative/explanatory texts to examine a topic and convey ideas and information clearly.
- W.5.2.B. Develop the topic with facts, definitions, concrete details, quotations, or other information and examples related to the topic.
- W.5.2.C. Link ideas within and across categories of information using words, phrases, and clauses (e.g., *in contrast, especially*).
- W.5.2.D. Use precise language and domain-specific vocabulary to inform about or explain the topic.
- W.5.2.E. Provide a concluding statement or section related to the information or explanation presented.

Continued

Table A2. (*continued*)

- 5.MD.C.5. Relate volume to the operations of multiplication and addition and solve real world and mathematical problems using volume.

- 5.MD.C.5.A. Find the volume of a right rectangular prism with whole-number side lengths by packing it with unit cubes, and show that the volume is the same as would be found by multiplying the edge lengths, equivalently by multiplying the height by the area of the base. Represent threefold whole-number products as volumes, e.g., to represent the associative property of multiplication.

- 5.MD.C.5.B. Apply the formulas $V = l \times w \times h$ and $V = b \times h$ for rectangular prisms to find volumes of right rectangular prisms with whole-number edge lengths in the context of solving real world and mathematical problems.

- 5.NBT.A.3.A. Read and write decimals to thousandths using base-ten numerals and number names.

- 5.NBT.A.4. Use place value understanding to round decimals to any place.

- 5.NBT.B.5. Fluently multiply multi-digit whole numbers using the standard algorithm.

- 5.NBT.B.7. Add, subtract, multiply, and divide decimals to hundredths, using concrete models or drawings and strategies based on place value, properties of operations, and/or the relationship between addition and subtraction; relate the strategy to a written method and explain the reasoning used.

- W.5.4. Produce clear and coherent writing in which the development and organization are appropriate to task, purpose, and audience.

- W.5.5. With guidance and support from peers and adults, develop and strengthen writing as needed by planning, revising, editing, rewriting, or trying a new approach.

- W.5.6. With some guidance and support from adults, use technology, including the internet, to produce and publish writing as well as to interact and collaborate with others; demonstrate sufficient command of keyboarding skills to type a minimum of two pages in a single sitting.

- W.5.7. Conduct short research projects that use several sources to build knowledge through investigation of different aspects of a topic.

- W.5.8. Recall relevant information from experiences or gather relevant information from print and digital sources; summarize or paraphrase information in notes and finished work, and provide a list of sources.

- W.5.9. Draw evidence from literary or informational texts to support analysis, reflection, and research.

SPEAKING AND LISTENING STANDARDS

- SL.5.1. Engage effectively in a range of collaborative discussions with diverse partners on *grade 5 topics and texts*, building on others' ideas and expressing their own clearly.

- SL.5.1.A. Come to discussions prepared, having read or studied required material; explicitly draw on that preparation and other information known about the topic to explore ideas under discussion.

- SL.5.1.B. Follow agreed-upon rules for discussions and carry out assigned roles.

- SL.5.1.C. Pose and respond to specific questions by making comments that contribute to the discussion and elaborate on the remarks of others.

- SL.5.1.D. Review key ideas expressed and draw conclusions in light of information and knowledge gained from the discussions.

Continued

Table A2. (*continued*)

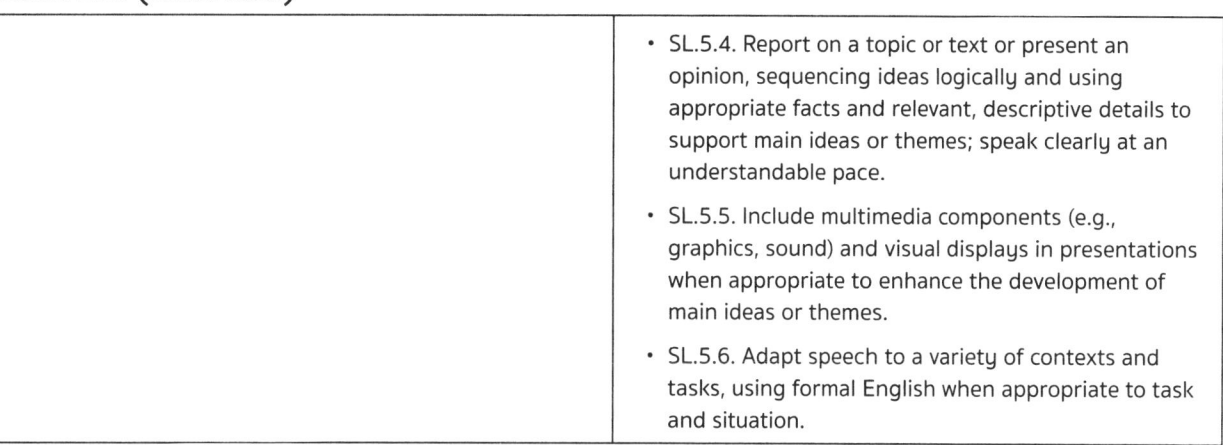

	• SL.5.4. Report on a topic or text or present an opinion, sequencing ideas logically and using appropriate facts and relevant, descriptive details to support main ideas or themes; speak clearly at an understandable pace.
	• SL.5.5. Include multimedia components (e.g., graphics, sound) and visual displays in presentations when appropriate to enhance the development of main ideas or themes.
	• SL.5.6. Adapt speech to a variety of contexts and tasks, using formal English when appropriate to task and situation.

Source: National Governors Association Center for Best Practices and Council of Chief State School Officers (NGAC and CCSSO). 2010. *Common core state standards.* Washington, DC: NGAC and CCSSO.

Table A3. 21st Century Skills From the Framework for 21st Century Learning

21st Century Skills	Learning Skills and Technology Tools	Teaching Strategies	Evidence of Success
INTERDISCIPLINARY THEMES • Health and Safety • Environmental Literacy • Science • Mathematics • Engineering Design Process (EDP)	• Use 21st century skills to understand issues surrounding the limited water supply. • Consider various needs when formulating solutions to a problem. • Apply understanding of science and mathematics concepts to create design solutions.	• Lessons include explorations of the potable water supply and the varying availability of fresh water across the United States. • Have students consider threats to the supply of fresh water in the United States. • Have students conduct research and analyze data related to various types of irrigation systems. • Have students perform calculations and apply their understanding of science concepts in their design projects.	• Students interpret, organize, and present information from activities and research in an effective format demonstrating understanding of irrigation and rainwater collection systems and the EDP.
LEARNING AND INNOVATION SKILLS • Creativity and Innovation • Critical Thinking and Problem Solving • Communication and Collaboration	• Use a variety of idea creation strategies to create, refine, and evaluate ideas both individually and as a member of a team. • Contribute productively to team work, remaining open to diverse ideas and opinions, assuming shared responsibility for team work, and valuing individual team member contributions. • Consider requirements and constraints in problem-solving and design, and demonstrate creativity and innovation in solutions. • Articulate thoughts and ideas effectively, using oral, written, and nonverbal communication skills in a variety of forms and contexts. • Listen effectively to decipher meaning, including knowledge, values, attitudes, and intentions. • Use communication for a variety of purposes (e.g., to inform, instruct, motivate, and persuade).	• Provide students with opportunities to brainstorm as teams, record notes about their ideas, and create sketches. • Scaffold students' team work in investigations and as they create models for their final challenge solution. • Have students work collaboratively as a class and in their teams to construct a public service message. • Model abstract mathematics and science concepts.	• Students are able to record their EDP thinking in multiple formats and can work in teams to share and improve on their ideas and designs. • Student ideas are documented and supported with evidence in their STEM Research Notebooks. • Design teams are able to communicate and collaborate as they interact in activities.

Continued

Table A3. (*continued*)

21st Century Skills	Learning Skills and Technology Tools	Teaching Strategies	Evidence of Success
INFORMATION, MEDIA, AND TECHNOLOGY SKILLS · Information Literacy · Media Literacy · ICT Literacy	· Evaluate information critically and competently. · Use information accurately and creatively for the issue or problem at hand. · Use technology as a tool to organize, evaluate, and communicate information.	· Have students use spreadsheets to organize and analyze data. · Students use technology to create presentations to share their final challenge solutions. · Have students consider a variety of media that can be used to publicly disseminate information.	· Students can use their research to guide decision making in the Rainwater Roundup Challenge. · Student slideshow presentations synthesize their learning and experiences throughout the module. · Students can use electronic spreadsheets to solve repetitive problems. · Student teams use media to present the class PSA message in an accurate and compelling manner.
LIFE AND CAREER SKILLS · Flexibility and Adaptability · Initiative and Self-Direction · Social and Cross-Cultural Skills · Productivity and Accountability · Leadership and Responsibility	· Adapt to varied roles, job responsibilities, schedules, and contexts. · Incorporate feedback effectively. · Balance tactical (short-term) and strategic (long-term) goals. · Utilize time and manage workload efficiently. · Define, prioritize, and monitor tasks without direct oversight. · Reflect critically on past experiences to inform future action. · Know when it is appropriate to listen and to speak. · Conduct themselves in a respectable, professional manner. · Leverage strengths of various team members to accomplish a common goal.	· Scaffold student understanding and teamwork competencies through a series of inquiry activities and topical research projects. · Facilitate students' and teams' self-monitoring by providing checkpoints for work. · Use EDP to encourage flexibility (through redesign), time management, and goal setting in structured group work. · Provide guidelines and practice opportunities for students to share their ideas and learning, emphasizing professional standards of behavior and inclusivity of all team members.	· Team projects are completed on time, with evidence of participation by all team members. · Teams' presentations include appropriate language and vocabulary. · Students are able to respond to questions regarding their designs and media messages.

Source: Partnership for 21st Century Learning. 2015. Framework for 21st Century Learning. *www.p21.org/our-work/p21-framework.*

Table A4. English Language Development Standards

ELD STANDARD 1: SOCIAL AND INSTRUCTIONAL LANGUAGE

English language learners communicate for Social and Instructional purposes within the school setting.

ELD STANDARD 2: THE LANGUAGE OF LANGUAGE ARTS

English language learners communicate information, ideas, and concepts necessary for academic success in the content area of Language Arts.

ELD STANDARD 3: THE LANGUAGE OF MATHEMATICS

English language learners communicate information, ideas, and concepts necessary for academic success in the content area of Mathematics.

ELD STANDARD 4: THE LANGUAGE OF SCIENCE

English language learners communicate information, ideas, and concepts necessary for academic success in the content area of Science.

ELD STANDARD 5: THE LANGUAGE OF SOCIAL STUDIES

English language learners communicate information, ideas, and concepts necessary for academic success in the content area of Social Studies.

Source: WIDA. 2012. 2012 amplification of the English language development standards: Kindergarten–grade 12, *https://wida.wisc.edu/teach/standards/eld.*

INDEX

Page numbers printed in **boldface type** indicate tables, figures, or handouts.

NATIONAL SCIENCE TEACHERS ASSOCIATION

INDEX